P9-DSV-772

Current Issues in the Economics of Water Resource Management

Economy & Environment

VOLUME 23

The titles published in this series are listed at the end of this volume.

Current Issues in the Economics of Water Resource Management

Theory, Applications and Policies

Edited by

Panos Pashardes
Professor of Economics,
Department of Economics,
University of Cyprus

Timothy Swanson
Professor of Law & Economics,
University College London,
United Kingdom

and

Anastasios Xepapadeas
Professor of Economics
Department of Economics,
University of Crete

KLUWER ACADEMIC PUBLISHERS
DORDRECHT / BOSTON / LONDON

A C.I.P. Catalogue record for this book is available from the Library of Congress.

ISBN 1-4020-0542-3

Published by Kluwer Academic Publishers,
P.O. Box 17, 3300 AA Dordrecht, The Netherlands.

Sold and distributed in North, Central and South America
by Kluwer Academic Publishers,
101 Philip Drive, Norwell, MA 02061, U.S.A.

In all other countries, sold and distributed
by Kluwer Academic Publishers,
P.O. Box 322, 3300 AH Dordrecht, The Netherlands.

Printed on acid-free paper

Printed in the Netherlands.

Table of Contents

Acknowledgements

We would like to thank the Planning Bureau, Government of Cyprus and the European Commission DG XII for financially supporting the research project. For sponsorship of the Symposium on 'Water Resource Management – Efficiency, Equity and Policy' that took place in Nicosia, September 22–24, 2000, we would like to thank the Cyprus Ministry of Agriculture, Natural Resources and the Environment, European Commission, DG XII, Joannou & Paraskevaides Ltd, Caramontanis Desalination Plants Ltd and George P. Zachariades Ltd. For hosting the Symposium we would like to thank the University of Cyprus, and for secretarial assistance Angela Shekersavva.

Panos Pashardes
Timothy Swanson
Anastasios Xepapadeas

List of Contributors

Bontemps, C., INRA, Economie et Sociologie Rurales, Chemin de Borde Rouge, B.P. 27, 31326 Castanet Tolosan Cedex, France

Casino, B., Department of Economic Analysis, University of Valencia, Avda. de los Naranjos, Edificio Departamental Oriental, 46022 Valencia, Spain

Couture, S., INRA, Economie et Sociologie Rurales, Chemin de Borde Rouge, B.P. 27, 31326 Castanet Tolosan Cedex, France

Garcia, S., CIRANO, 2020 University, 24e etage, Bureau 2406, Montreal, Quebec, Canada H3A 2A5

Groom, B., Department of Economics, University College London, Gower Street, London WC1E 6BT, U.K.

Hellegers, P.J.G.J., Agricultural Economics Research Institute (LEI), P.O. Box 29703, 2502 LS The Hague, The Netherlands

Karagiannis, G., Department of Economics, University of Ioannina, 45110 Ioannina, Greece

Koundouri, P., Department of Economics, School of Business, The University of Reading, Reading RG6 6AA, U.K.; and Department of Economics & CSERGE/Economics, University College London, Gower Street, London WC1E 6BT, U.K.

Martínez-Espiñeira, R., Department of Economics, St. Francis Xavier University, P.O. Box 5000, Antigonish, Nova Scotia, Canada B2G 2W5

Nauges, C., LEERNA-INRA, Manufacture des Tabacs, 21, Allée de Brienne, 31000 Toulouse, France

Pashardes, P., Department of Economics, University of Cyprus, P.O. Box 20537, 1678 Nicosia, Cyprus

Roseta Palma, C., DEGEI, University of Aveiro, Campus de Santiago, 3810-193 Aveiro, Portugal

Rubio, S.J., Department of Economic Analysis, University of Valencia, Avda. de los Naranjos, Edificio Departamental Oriental, 46022 Valencia, Spain

Swanson, T., Department of Economics and Faculty of Laws, University College

London, Gower Street, London WC1E 6BT, U.K.

Thomas, A., LEERNA-INRA, Manufacture des Tabacs, 21, Allée de Brienne, 31000 Toulouse, France

Timmins, C., Department of Economics, Yale University, P.O. Box 208264, New Haven, CT 06520-8264, U.S.A.

Tzouvelekas, V., Department of Economics, University of Crete, 74100 Rethymnon, Crete, Greece

Van Ierland, E.C., Environmental Economics and Natural Resources Group, Wageningen University, P.O. Box 8130, 6700 EW Wageningen, The Netherlands

Xepapadeas, A., Department of Economics, University of Crete, 74100 Rethymnon, Crete, Greece

PART 1: INTRODUCTION

Current Issues in the Economics of Water Resource Management: Introduction

PANOS PASHARDES*, TIMOTHY SWANSON and
ANASTASIOS XEPAPADEAS

1. Introduction

Many economists have recently discovered that the problem of water resource management is an important and interesting area for the application of the tools of economic theory and econometrics. At the same time many water specialists from other disciplines have discovered that the language and tools of economics are helpful in furthering the understanding of water management problems. For these reasons this branch of economics has developed rapidly in recent decades and is likely to continue to do so.

Despite having made a major contribution to the investigation of water management issues over the past few decades, in many ways the application of economics in the area is still in its infancy, both in terms of theory and practice. Take, for example, the use of price as a water consumption management tool. This has been an issue of growing concern in Europe and the United States where private and public utilities have to assess the welfare implications of various pricing schemes. Economists have attempted to shed some light on the matter through water demand analysis. However, very rarely is water demand studied within a theoretical framework consistent with the fundamental principles of consumer theory, such as additivity, price homogeneity, and symmetry. Furthermore, any results obtained are difficult to interpret in terms of social welfare, because water demand is seldom studied within the context of a model based on a well-defined utility function. Developing the welfare economics fundamentals of water management is one of the key steps being taken today in this literature.

Even while the fundamentals of demand analysis are being resolved, it is apparent that only slow progress has been made in implementing the results in the form of coherent and practicable water management policies. In part this is due to the complexity of issues surrounding water management. Indeed, very few (if any) commodities possess as many idiosyncratic characteristics; water is a multi-faceted

* We thank the Government of Cyprus and the European Commission, DG XII, for financial support. The responsibility for errors is, of course, ours.

P. Pashardes et al. (eds.),
Current Issues in the Economics of Water Resource Management, 3–21.

good in time, space and consumer preferences. There are also few resources that attract as much competition among its potential users, as does water in the arid regions. Historical rights intersect with modern requirements to generate hopeless stalemates in water allocation. In these circumstances, not many water analysts can be convinced that the standard economic paradigm provides an adequate theoretical framework for the analysis of water related policies. Too much complexity is present in the usual water management situation for simplistic solutions to suffice.

The scepticism about the adequacy of economic theory to provide answers is reinforced by the absence of the information required for analysis to reach practical policy conclusions. The lack of good quality data is a fundamental problem in this area. Reliable data – both about the water characteristics themselves and about the forces driving agents' behaviour – are absolutely critical to the development of sound economic models. Yet, the lack of data is most prevalent where it matters most, in arid regions, because these also tend to be in some of the least developed regions of the world.

Taking a pessimistic view one can have serious doubts about the usefulness of applying the tools of modern economics to the complicated problems of water resources. We do not subscribe to this negative view. The fact that many economic tools taken off-the-shelf may not be appropriate for the study of issues pertaining to water resources, does not mean that these tools are not amenable to adaptation for this purpose. Indeed, much of the challenge that lies ahead is to find ways to adapt rigorous economic tools to this important task.

This volume is a step in this direction. It contains 12 papers, each attempting to employ economic tools to answer questions about water management. Furthermore, the majority of the papers use econometric techniques to investigate these questions on empirical grounds.[1] In this introduction we discuss these papers in three sections. Section 2 concerns the analysis of demand for water and considers four papers in two subsections, one focusing on residential and one on irrigation demand. Section 3 examines the efficiency aspects of water use: the two papers discussed in the first subsection are about the measurement of water efficiency and the two discussed in the second subsection are about efficiency in terms of strategic use and management of water resources. Section 4 concentrates on water management policy and discusses the problems involved in applying economic tools to complex goods. Section 5 concludes this introduction.

2. Demand for Water

The study of demand for water is considered to be part of a strategy for the management of water resources in the sense that it provides information about the effects of control variables on water use. In particular, knowledge of how price influence water-concerned activity can be a crucial ingredient in the design of a water pricing policy aimed at keeping the use of water at a level balancing water supply.

Water is demanded for final consumption and as an intermediate input in production. Residential water demand accounts for most of the final water consumption whereas irrigation accounts for most of intermediate water demand, especially in water-scarce areas. Four papers in this volume are about water demand: the first two address issues concerning residential water demand, the third examines how irrigation water demand can be estimated using programming methods and the fourth considers the effect of an externality associated with water demanded for irrigation.

Consumer theory provides a useful and convenient framework for residential water demand analysis and for the investigation of the efficiency and distributional effects of alternative water pricing systems. It is, therefore, not surprising that so many investigators use consumer demand analysis tools to estimate the effects of price on water demand. The findings of some of these studies (mostly conducted in the United States and, more recently, also in Europe) are summarised by Herrington (1987). Overall the existing empirical results suggest that the price elasticity of demand for water is significantly different from zero and lie somewhere under −0.3.

A peculiar feature of the water consumption data that requires attention in modelling consumer behaviour arises from the fact that most countries adopt an increasing block pricing (tariff) structure. In general, increasing block (progressive) tariffs are becoming more common in developed as well as developing economies because this pricing system conforms to the perception that it can be used as a tool for both social justice and conservation of a scarce natural resource. Strictly speaking, there are no obvious efficiency arguments for an increasing block water tariff structure and the evidence in favour of the argument that this pricing structure has a 'psychological' effect helping water demand management is ambiguous: OECD (1987) reports evidence in favour of this argument in Japan, Italy, Denmark and Switzerland (Zurich) while other studies (e.g. the metering trials in the Thames and Yorkshire water authorities in the UK) prove inconclusive.

In the context of consumer theory, an increasing tariff structure results in a nonlinear budget constraint. This complicates the conventional approach to modelling consumer behaviour based on utility maximisation or cost minimisation. Hausman (1978) formalised the economic theory of consumer optimisation under a nonlinear budget constraint in the context of labour supply. The practical conclusion coming from Hausman's work is that the segment (block price) along the budget constraint where the consumer reaches the highest indifference curve is endogenous. Therefore, the quantity as well as the (block) price must be estimated simultaneously. Authors using Hausman's approach to solve the problem of increasing tariff structure in demand for water include Hewitt and Hanemann (1995), Coral et al. (1998) and Pint (1999).[2]

An alternative approach to tackling the increasing tariff structure in water demand analysis originates from the work of Taylor (1975) and Nordin (1976). Taylor (1975) tried to account for the increasing tariff structure in the context of demand for electricity by including both the marginal and average price among the explanatory variables. Subsequently Nordin (1976) demonstrated that Taylor's specification should be modified to include a 'difference' variable to account for the distorted income effect imposed by the increasing tariff structure. In the case of multiple tariffs (and those cases where fixed quota and/or a free allowance is used) the difference variable is the difference between the total bill and what the user would have paid if all units were charged at the marginal price. Empirical studies of residential demand for water using either Nordin's difference variable on its own or in combination with Taylor's marginal price in their empirical models include Billings (1982), Schefter and David (1985) and Martinez-Espineira (2000, and in this volume).

2.1. Residential Demand for Water

The first paper in this volume by Martinez-Espineira adopts the Taylor–Nordin approach to estimating a domestic water demand function using quarterly billing data at the municipal level from the Northwest of Spain. The main focus of the paper is the calculation of the price elasticity of demand. The author argues that by using monthly (as opposed to annual) data is important because the seasonal dimension of water demand can also be more investigated.

The main objective of the Martinez-Espineira's paper is the analysis of the marginal price elasticity and the relation between the coefficients of income and Nordin's difference variable. The innovative feature of this study is the use of the proportion of water users falling in each block of consumption in the empirical specification. This is argued to be theoretically correct when analysing demands under block tariffs with aggregate data because it allows the marginal price and difference variables to be defined correctly. The results obtained from the proposed empirical specification are also compared to those obtained from applying a conventional price specification to the same data set. Yet, the innovation introduced by Martinez-Espineira in the definition of the marginal price and difference variables does not help accept that the income and difference variable have the same but opposite sign effect on demand, as implied by the theory. This is common to nearly all known studies of water demand using the Taylor–Nordin approach. In our view the author is probably right in saying that Nordin's solution to the increasing block pricing problem may not after all be appropriate for the study of water demand. The Hausman (1978) approach mentioned above is theoretically superior to the Taylor–Nordin approach and does not impose parameter restrictions on the data other than those pertaining to the fundamentals of consumer theory.

The marginal price elasticities estimated by Martinez-Espineira conforms to expectation. The price specification that accounts for the changing proportion of water users in each block yields a higher elasticity (−0.47) compared to the specification ignoring this feature of the data. However, this difference is not found to be statistically significant, a result attributed to the low power of the test (small sample size limiting the accuracy of estimates). In conclusion, the paper provides a theoretically correct price specification for demand functions under block pricing and aggregate data. The empirical findings in the paper, however, are not conclusive and further empirical work using more data and alternative (nonlinear) demand functions, is needed to show the practical implications of the arguments put forward by the Martinez-Espineira's paper.

Static empirical consumer demand functions estimated with aggregate data are well known to suffer form serial correlation and other statistical problems associated with misspecified dynamics. These dynamics arise because consumers do not react immediately to a change in prices due to their largely predetermined lifestyle. In the case of demand for water, for example, current purchases can be largely predetermined due to commitments arising from past purchases such as swimming pools, bathtubs, dishwashing machines, etc. Muellbauer and Pashardes (1992) show that the autoregressive nature of consumer demand data can be captured in a theoretically consistent manner by incorporating intertemporal aspects of consumer behaviour in the model through habit formation and durability. This entails the use of lagged dependent variable effects in the model, subject to restrictions required for the adding-up, symmetry and homogeneity properties of the model to hold at both the intratemporal and intertemporal levels.

Most of water demand studies estimated with aggregate data are conducted in a static framework, i.e. current water consumption is modelled as a function of current price, income and, possibly, other variables.[3] The paper by Nauges and Thomas breaks away from this tradition and proposes a dynamic model which explicitly takes into account the possibility that current water can be influenced by past water use (through habits or stocks of durables). It is argued that the low price charged for water in many industrialised countries over the years has helped built habits in consumption; and that this can be responsible for the slow consumer response to the sharp rises in the price of water introduced by the French government in 1992.

The nonlinear dynamic model used in the Nauges and Thomas paper is derived from an intertemporal structural optimisation program solved by the municipality to maximize local consumers' utility and reduce the debt of its water service. The control variable is the price of water charged to the domestic users and is used by the local authority to determine the consumption path. Under assumptions describing the annual price adjustment and community heterogeneity current consumption is written as a linear function of lagged consumption, current price and income, where the error term combines time-invariant individual heterogeneity and a multiplicative time-varying component. The authors then propose an estimator combining first- and quasi-difference transformations and show that this is consistent and efficient in the case of the mixed error term structure of their model.

In their empirical analysis Nauges and Thomas estimate both a non linear model directly derived from the optimisation program and a linear model with the mixed error structure. They show that the two models are asymptotically equivalent. The estimation is conducted on a sample of French communities observed during the 1988–1993 period and yields a long run price elasticity of −0.4 and a short run price elasticity of −0.26. The higher, in absolute terms, long run elasticity reflects consumer habits slowing down the adaptation of water demand to price variation. The paper concludes by urging local authorities to use the long run price elasticity of demand for water in assessing the impact of alternative tariff policies on the level of consumption and/or on consumer welfare.

2.2. Irrigation Water Demand

Water demand for irrigation accounts for the lion's share of water use in water-scarce areas. Farmers' response to changes in the price of water is normally estimated using either an econometric or a programming approach. The econometric approach assumes that farmers behave as a profit maximising firm taking decisions concerning crop choices, crop-level allocations of land and water use (Moore and Negri, 1992; Moore et al., 1994; Hassine and Thomas, 1997). The programming approach derives water demand relationships from 'shadow prices' obtained by computer simulations of profit maximising behaviour (Shunway, 1973; Howitt et al., 1980).

In their paper Bontemps and Couture opt for a programming method based on the evaluation of the farmer's value for water in France. The farmer's value for water is defined as the maximum amount of money the farmer would be willing to pay for the use of the resource. A crop-growth simulation model is used in place of a production function in the estimation, in order to obtain precise results concerning water-yield relations. The model is estimated in two stages. The first stage uses a numeric optimisation algorithm to evaluate profit maximising allocation of a limited water supply over an irrigation season. This generates a database consisting of levels of quotas and associated maximised profits. At the second stage nonparametric methods are applied to the database to estimate the maximised profit functions.

The results obtained from applying this estimation method to water irrigation data in South-West France show that the irrigation water demand functions are decreasing, nonlinear and strongly dependent on weather conditions. Water demand for irrigation appears inelastic at low water prices and elastic at high prices. The price levels at which water demand appears elastic depend on the climatic conditions and vary from around 0.30 F/m^3 for wet weather conditions up to 1.60 F/m^3

for a dry year. Thus, the work of Bontemps and Couture in this volume suggests that the climate and the price at which the elasticity is evaluated are both important factors in defining appropriate tariff structure to curb irrigation water demand.

The last paper in this section is not exactly demand for water but demand for agricultural land. The importance of water demand in this paper relates to the fact that scarcity has a qualitative dimension that affects the productivity of land as an input in agricultural production. Koundouri and Pashardes argue that this environmentally determined productivity differentials can be reflected in the structure of land rents and prices. Therefore, using data on land rent or land value for different properties, they try to estimate the effect which fresh groundwater quality has on the value of (willingness to pay for) the land. Furthermore, it is argued that failure to account for alternative uses of land as an input in production (e.g. tourism) can give rise to a sample selection problem (Hausman et al., 1992) resulting in parameter estimates reflecting misleading shadow prices of land and irrigation water quality.

This is not the first paper to investigate the relationship between land prices and water access in terms of quantity and quality. King and Sinden (1988), Caswell and Zilberman (1986) and Torell et al. (1990), among others, use hedonic techniques to do so while travel cost techniques have been used to measure the welfare effects to changes in water quality of recreational sites by, for example, Caulkins et al. (1986), Smith and Desvousges (1986) and Bockstael et al. (1987). The novelty in this paper is that the option to purchase land is allowed to be endogenous to the price paid, as it is the decision how much land to purchase. Land close to the seaside is demanded for use as an input either in agricultural production or in tourist development. The proximity to the sea decreases the probability of land usage for agriculture due to salination of groundwater supplies and increases the probability of tourism usage due to attractiveness to tourists. Yet, land parcels closed to the seaside may continue to be used in agriculture (in spite of the poor quality of their underwater supplies) because they are still below the tourist development reservation price. In this context water quality can determine whether a plot of land should be included in or excluded from the sample of agricultural land.

Koundouri and Pashardes show how this selectivity problem caused by water salinity affects the hedonic valuation of agricultural land. The empirical analysis, based on data collected from surveying 282 owners of land parcels in the Kiti area of Cyprus, supports the hypothesis that failing to correct for sample selection results in a biased valuation of groundwater salinity as an attribute of agricultural land. The estimated marginal willingness to pay for fresh groundwater in a model correcting for sample selection is statistically not different from zero, whereas without this correction this value appears to have a significant positive effect on the value of land. The authors argue that without the selectivity correction one ignores the fact that the cost of lower groundwater quality can be offset by an increasing probability of switching to the more lucrative tourism industry.

The ideas expressed in the Koundouri and Pashardes paper have wider implications for the applications of hedonic techniques to derive willingness to pay for environmental and/or resource quality. This paper shows that water salinity has a negative effect on the value of land demanded for agricultural but not tourist use. There might be other situations where alternative uses of environmental resources do not all require the same attributes. In these situations one should question the sample selection criteria when using hedonic techniques for the valuation of these attributes to avoid biased conclusions.

3. Water Efficiency

The concept of efficiency in the use of water is directly associated with the ability of water systems, either surface or groundwater, to support in a sustainable way the needs of humans as well as of natural ecosystems. In this book the concept of efficiency is approached along two different lines: the first relates to the concept of the technically efficient use of irrigation water as an input in agricultural production, while the second relates to the concept of efficient management of water resources and the design of effective policies.

Efficiency in production and the measurement of productive efficiency is based on the seminal work of Koopmans (1951) Debreu (1951), Farrell (1957). Radial measures of technical efficiency focus either on the maximal equiproportional reduction of all variable inputs or the maximal equiproportional expansion of all outputs. On the other hand allocative efficiency relates to the employment of the *right* proportions of inputs *given* their prices. Farrell defines overall productive efficiency as the product of technical and allocative efficiency.

Regarding irrigation water, while allocative efficiency relates to the pricing of irrigation water, technical efficiency has been traditionally defined by the ratio of effective water use, i.e., the amount of water actually utilised by crop, to the water applied to the crop. In this way technical efficiency is defined in terms of the underlying irrigation technology in engineering terms.[4] Thus technical efficiency is defined in terms of the technical characteristics of the irrigation technology. On the other hand technical inefficiencies, in the sense of not attaining – for given levels of the inputs – the *minimum* feasible water use, in order to obtain given output, can be observed among farmers that use the *same* irrigation technology. This is input-specific technical inefficiency associated with management failures. These types of inefficiencies might result in waste of water resources and their identification could provide some policy guidelines for remedying such sources of inefficiencies.

Efficient management of water resources is associated with the potential need for public intervention in water resource allocation. When water markets are missing or are thin, when there are social costs or benefits associated with the use of water resources which are not included in the private agents' objective functions, or when water resources have open access characteristics, then public intervention may allocate water resources more efficiently. Intervention could take the form of regulation (market-based instruments or command-and-control), public investment or public ownership and regulation (Young and Haveman, 1985).

An approach to this problem, which is common in environmental and resource economics management, is to define first a solution to the water management problem associated with a social planner or regulator. The social planner maximizes a social welfare function appropriately defined to include any externalities relating either to the use of water (flow effects) or to the services provided by the existing stock of water, subject to the natural constraints of the hydrological system. For example, the use of groundwater for agriculture relates to agricultural run-off externalities that reduce the quality of groundwater due to intrusion of pollutants such as nitrate or pesticides. Stock services on the other hand, can be associated with stock effects on pumping costs, buffer values, option values, recreation values or ecological values. Some or all of these externalities are ignored by private agents choosing their water use in order to maximize private profits. As a result the outcome of private markets is inferior on welfare grounds relative to the social optimum. Furthermore different assumptions regarding the information structure of the private agents, especially open loop *versus* feedback[5] in dynamic setups of water management, affect the size of inefficiencies associated with the privately optimal solution.

The efficient use of water resources can be attained by the choice of an appropriate regulatory scheme that seeks to induce private agents to use water resources in a way that approximates the socially-optimal use. Thus, one part of the efficient management issue is to identify deviations between the social optimum and the private optimum. The second is to design and implement an efficient policy, that is a policy that achieves – or at least approximates – at a minimum cost the targets set by social welfare maximisation.

3.1. Measurement of Water Efficiency

In their paper Karagiannis, Tzouvelekas and Xepapadeas explore the issue of irrigation water efficiency by introducing the concept of input specific technical efficiency. This concept contrasts with measures previously used in the literature such as output-oriented and input-oriented measures of technical efficiency incapable of identifying the efficient use of individual inputs. The proposed measure is a nonradial input-oriented measure of input-specific technical efficiency (Kopp, 1981). This measure has an economic rather than an engineering meaning since it relates the maximum feasible water use to the observed water use, given the irrigation technology and the observed levels of output and other inputs used. Thus, the proposed measure of technical efficiency has an input-conserving interpretation but is not associated with a cost saving concept since it is a nonradial measure. On the other hand, its major advantage is that it reveals inefficiencies in water use for any given irrigation technology, such as drip or sprinkler technologies, which are associated with the managerial capabilities of farmers rather than the water efficiency technology potential of the specific irrigation technology.

A stochastic production frontier approach based on the inefficiency effect model (Battese and Coelli, 1995) is used to obtain estimates at the farm level of both technical and irrigation water efficiency, for a randomly selected sample of 50 out-of-season vegetable growing firms located in Crete, Greece. In addition a second stage approach is used to identify the factors affecting water efficiency differentials across farms.

Empirical results indicate that irrigation water efficiency is on average much lower than technical efficiency, implying that significant savings are possible by more efficient use of irrigation water, given the present state of technology and inputs use. This finding does not support previous beliefs that advanced irrigation practices (i.e., drip irrigation) would, by definition, improve the efficient use of irrigation water and reduce waste of natural resources. The results obtained by Karagiannis, Tzouvelekas and Xepapadeas indicate that even farmers utilising modern technologies are unable to fully exploit their water-saving potential. Policy implications stemming from this paper suggest that education and extension are among the main factors favorably affecting the degree of irrigation water efficiency for a given technology, indicating therefore that significant savings in water can be achieved through the development of better water management practices. Although the pricing of irrigation water seems to be at the core of water conservation issues, the empirical findings of this paper suggest that the appropriate use of irrigation systems could prove to be an equally important issue in implementing better water management.

3.2. Efficient Use and Management of Water

The first paper in this section by Rubio and Casino analyses the issue of the pumping cost externality in relation to the strategic externality in common property groundwater extraction. As has been shown by Negri (1989), in a differential game framework, a feedback solution – assuming existence and uniqueness of such a solution – that captures both the pumping cost and the strategic externality increases inefficiencies in groundwater management, relative to the open loop solution that captures only the pumping cost externality. A similar result is also obtained by Provencher and Burt (1993) under the assumption of a concave value function. However, another line of approach (Tsutsui and Mino, 1990; Dockner and Van Long, 1993), suggests that in feedback solutions with *nonlinear strategies* inefficiencies could be reduced, even to the extent that cooperative outcomes could be supported by nonlinear feedback strategies.

Rubio and Casino analyze an important and timely issue in differential games, namely the existence and the properties of nonlinear feedback equilibrium strategies in the context of open access groundwater extraction. The results they obtained suggest that the inefficiency reducing effect of nonlinear strategies requires an unrestricted domain for the initial valuc of the state variable. This is not however the case for groundwater management, where the initial value is restricted and corresponds to the natural hydrological steady state. Thus, only the *global* linear strategy leads to a steady state for the Markov feedback equilibrium, which is inferior to the open loop equilibrium, while both the open loop and the feedback equilibria are inferior relative to the socially optimal solution. Furthermore, Rubio and Casino find that when the storage capacity of the aquifer increases, the difference between the socially-optimal exploitation and the privately-optimal exploitation decreases, confirming thus the Gisser–Sanchez rule.

The second paper by Timmins, analyzes the overall effectiveness, in dealing with water scarcity problems, of policies based on nonprice-demand-side technology standards, relative to efficient water pricing policies. Timmins suggests that in the short-run a mandatory policy of low-flow appliance installation, such as low flow toilets or shower heads, may mitigate water scarcity, even when during the application of such a policy water is priced below the cost of provision at the margin; however long run effects might offset any short run conservation gains. The reason is that the short-run gains from low-flow appliance installation may allow the water authority to pursue an even more aggressive underpricing policy, on equity or political grounds, so that in the long run actual water conservation is minimal.

Timmins develops a model that explains the decisions of municipal water managers to underprice water relative to static marginal cost of provision even in the presence of dynamic cost associated with depletion effects of the aquifer's water stock. The parameters of the municipal manager's objective function, which is consistent with underpricing behavior, are recovered econometrically. The estimated objective function is used to explore the short and long-run conservation potential of a policy of nonprice-demand-side technology standards and to compare such a policy to alternative pricing regimes. It is shown that price policy is cost-effective in attaining a particular level of groundwater extraction. If, on the other hand nonprice policy is preferable on equity grounds, simulations indicate that long-run price reducing reaction of managers almost eliminates all short-run water conservation gains of nonprice policies. Timmins' paper provides a method for the econometric estimation of the parameters of the managers' objective, and succeeds in uncovering price effects that might undermine any advantages of commonly used water conservation policies, indicating thus ways for effective policy design by combining nonprice demand management policies with some pricing austerity.

4. Water Management Policies

This section of the volume covers a wide range of topics concerned with water management, ranging from the basic instruments required for water allocation to the niceties of utility regulation under imperfect information. The range of topics considered indicates the context-dependent nature of water management policy. The one salient feature of water management is that it concerns the management of all of the various actors and activities that are hydrologically linked. The nature of these interactions may vary significantly. For example, agricultural users may interact with one another in their extraction from a common aquifer, or they might interact with consumers of groundwater for drinking purposes by means of their application of pesticides within the watershed.

Irrespective of the nature of the use, the economic objective of water management policy is the same – to internalise the social cost of water allocations to the respective users. The basic economics of water management policy is well-developed and exposited (Neher, 1990). The problem is to develop and to apply instruments that are capable of achieving this object within the myriad of contexts in which these externalities arise. Such policies have been generally discussed and developed in the common aquifer case (Provencher and Burt, 1993). Also they have been discussed in the context of nonpoint source pollution (Griffin and Bromley, 1982; Xepapadeas, 1992). They have also been discussed in the context of conjunctive (groundwater/surface water) users (Provencher, 1995).

The instruments that are usually discussed are either of the nature of *water pricing* (see, e.g., Feinermann and Knapp, 1983) or *property rights* (Provencher, 1993) sorts of systems. Both approaches have been used for the allocation of water, and sometimes within the same jurisdiction (Zilberman et al., 1997). These instruments are intended to give effect to the social cost of water allocations, and thereby internalise that opportunity cost to each and every user of the water resource. Water pricing does so by attempting to set a marginal social value of water use, and then allows water users to determine individual use by reference to this price. Property rights approaches attempt to complete the markets in water, and then allow water users to transfer water rights between them until the socially optimal allocation of water results.

Of course, given the context-dependence of the problem, the policy problem is a case of the 'devil being in the details'. For example, how is the cost of water allocations to nonpoint pollution to be internalised under either system? Does the cost of water use (contamination) get charged to the farmer applying the pesticide, the chemical company producing it, either or both? If property rights to groundwater are created, does the farmer need to acquire rights before applying the pesticide, or does the producer need to acquire rights before selling it? How many such property rights need be acquired, given the nonexclusive nature of the use? Does this indicate that the regulator needs to create property rights in the many facets of the water resource (quantity, quality, present, future)? If so, how does the system incorporate the interactions between these various facets? Clearly, this one example suffices to indicate the complexity of the policy problem that we address here.

It also begins to point to the fact that there are both vertical and horizontal relationships implicit in water policy considerations. Just as the above example brought into consideration both the producer of the chemical contaminant and its ultimate users, it is demonstrated that the vertical relationships within which water is used is as important as the horizontal ones (Swanson and Vighi, 1999). Industries (such as the agro-chemical industry) make important decisions about water management, and the study of water management must extend across these industries as well as across the whole of the watershed. This is another dimension of complexity involved in the problem. It indicates the importance of the regulator's role in industrial regulation regarding water use and allocation, as well as its role in managing individual users. The clearest example of a vertical relationship with clear implications for water management is the water supply industry itself, and this is a critical element of the water management problem (Thomas, 1995).

In this part of the volume we provide examples of the policy problems that arise from considering applications of economic policy prescriptions within specific contexts. The four papers together represent a wide range of the issues in the area, but this general discussion indicates that they still represent only the tip of the iceberg.

4.1. Water as a Multifaceted Resource – Horizontal Management Considerations

The problem of water policies that will optimally allocate water between its many competing users implies the management of the many different facets of the resource, as set out above. In the absence of an adequate number of facets within the policy mechanism, the match between policy and problem will be imperfect. The first two papers in this section illustrate this problem.

The paper by Catarina Roseta Palma develops the analytics of the multi-faceted resource in the context of groundwater users and a common aquifer. Her model very explicitly pursues the problem of distinct facets of resource, in terms of the quantity and quality of water. Her example is that of the farmers that apply groundwater to their previously chemically-treated lands, thereby generating a problem of chemical contamination of the aquifer at the same time that they are extracting the resource. Water applied to agricultural lands will react with the previously applied chemicals within the soil, and then leach through the subsoil and into the water table. In this manner, farmers overlying an aquifer are consuming two facets of the resource at once.

Palma develops the problem within a model of resource usage with two stock variables (the two facets of the resource) in which it is initially assumed that both stocks are relevant to the user's production decision. She first demonstrates that the common pool problem will result in the additional externality (in addition to the usual pumping cost) concerning the failure to consider adequately the cost of chemical contamination in groundwater usage. She then demonstrates that a single regulator (of either quantity or quality) will be unable to resolve the problem, since a single instrument cannot address a multi-faceted problem. The same will be true if the quality consideration arises not from production costs but from environmental quality constraints. The efficient solution cannot be achieved unless the multi-faceted nature of the resource is taken into consideration in fashioning it.

The paper by Groom and Swanson is another simple evocation of this particular policy problem. In that paper the problem concerns the relationship between groundwater and surface water users concerning their respective preferred times of use. The relationship is assumed to be a conjunctive one, i.e. the groundwater users have the option of making use of either ground or surface waters. This implies that the two groups are linked by their surface water usage. The problem analysed by Groom and Swanson concerns the differences between the two groups regarding their preferred time horizon of groundwater usage. If, for example, the group that has access to the groundwater is distinct from the other in important ways, e.g. agricultural rather than residential users, then it might be the case that they have very different preferred time horizons of water usage. Agricultural users might be larger entities with more risk smoothing capabilities, for example, and hence they may be relatively risk-neutral.

In the paper the two groups are assumed to be distinguished by their relative aversion to the stochasticity of arriving surface water flows, and this determines their preferred timing of groundwater usage within the watershed. The policy relevant point of the paper is that even well-defined and exclusive property rights in the groundwater would be ineffective in achieving the optimal allocation if they are not marketable between the distinct groups. This is indicative of the fact that the time-horizon facet of the good must not only be well-defined, but also tradeable in order for the property right mechanism to be effective in achieving the efficient solution. The conclusion of the paper is that, when an existing institution (here, the property rights system) does not enable the achievement of the efficient solution, then new instruments might be devised by the parties in order to pursue the same purpose. In this context, the surface water users are forced to invest in the creation of other mechanisms for smoothing water flows, since the institutional apparatus is not adequate for this purpose, hence the reference to redundant reservoirs in the title. Ineffective policy measures can result in inefficient capital investments.

The two papers together illustrate the nature of the water management problem: the multi-faceted relationships between users imbedded within the water system generate complicated policy problems. The abstract nature of the policy solution is straightforward, but the details introduce complexities requiring larger numbers of instruments and the recognition of the interaction between them.

4.2. Water as an Industrial Management Problem – Vertical Considerations

As the water resource becomes increasingly affected by industrial choices, the role of the regulator becomes more concerned with managing firms rather than individuals. This involves the management of industries that indirectly use or impact water supplies, but increasingly it involves the management of industries that directly impact water supplies. The commercial supply of water is becoming increasingly common in various parts of the world.

In this context the regulator's problem often concerns the motivation of individual firms to pursue optimal investments (in some facet of water quality) within an environment of imperfect information. That is, even if the regulator is able to perceive and regulate the price schedules under which firms supply water, this is inadequate to achieve the social objective if other facets (water quality, infrastructural investments, etc.) are not regulated. Once again the problem concerns the multi-faceted nature of the good, and in this circumstance the ability of the regulator to motivate the firms to pursue objectives related to multiple facets (price, quality).

The paper by Garcia and Thomas pursues these issues in the context of the problem of motivating private water supply firms to pursue investments in water transmission infrastructure. Assuming that most of this infrastructure lies underground, it is then literally invisible. Then the information on the quantity of water that must be delivered in order to achieve a given level of water consumption may be private to the firm. If the society cares about the level of 'wastage' within the system, then it will be necessary to regulate the firms for the level of their investments in infrastructural quality as well as for their pricing behaviour, yet it may not be possible to view their individual cost-efficiency (here on account of accounting methods).

Garcia and Thomas develop a model in which such firms (of unknown cost efficiency) may nevertheless be regulated by means of the construction of an optimal menu of 'contracts' combining the characteristics of: consumed water per customer, transmission wastage and financial transfers. By offering a menu of such contracts the individual firms identify their level of efficiency through their selection of contract, and the efficient level of investment by individual firms may be achieved. They illustrate the application of the regulatory method in an application to the private commercial water supply sector in France.

Of course, the presence of imperfect information does not allow for the pursuit of the first-best policy solution. Each firm will be directed toward the pursuit of its individually optimal level of investment in quality; however, the presence of uncertainty always generates an informational rent for the regulated entity. The regulator is creating instruments that are necessarily second-best.

This case study demonstrates some of the same points made previously. The management of the water resource involves the pursuit of multiple objectives, here the simultaneous regulation of quantity and quality. Given the number of users and complexity of their interactions, it is clear that the regulatory environment will require multiple instruments (or multi-faceted ones as in the contracts offered here), and that the results will nevertheless occur within the world of the second best.

4.3. Water Policy and Water Practice – Second Best

Even if water policy is complicated in theory and solutions remain second-best in nature, the actual practice of water management remains even further behind the considerations of policy. Few (if any) systems exist that use full-cost water pricing methodologies or fully transferable property rights. The problems in water management concern not just the complexities of water policy in principle, but the problems concerning how to put these prescriptions into practice.

The paper by Hellegers and van Ierland illustrates this point in the case of the Netherlands. The practice of water extraction for irrigation is now priced, but only to those users of more than 30,000 cubic meters per annum, i.e. 1–2% of all irrigators within the country. There are also property rights to extract water, but these are historical rights and nontransferable. This also prevents the consideration of more recently recognised and valued uses of water, such as environmental allocations. The only other allocation measures currently in use are some seasonal bans on irrigation and some dissemination of information about efficient timing of irrigation.

The paper illustrates just how far practice and policy are from the optimal. For some reason the prescriptions of economists are not being realised in the field, even in relatively advanced countries such as the Netherlands. This can be partly the result of political economy and partly the result of path dependence.

5. Conclusion

Water management is a topic that continues to fascinate and perplex the economic analyst. It is a familiar producer and consumer good that most of us make use of every day, and yet it is also a complex and multi-faceted commodity that frustrates all attempts to simplify it. The complexity of the substance – its wide range of facets and systemic effects – renders simple economic analysis useless. Despite its ubiquitous and transparent nature, it remains opaque and unyielding when analysed in simplistic fashion.

This volume contains a large number of studies that demonstrate this point, and also demonstrate a wide range of approaches to dealing with this complexity. The first section of the volume presented a set of papers dealing with the frontiers of demand estimation regarding water utilisation. The second set of papers dealt with the complex issues of efficiency and externality in regard to a complex good. The third set of papers addressed the interface between the governance/management structure and the resource. In all three cases, the water resource was treated with the complexity that it deserves, and the papers demonstrate the small amounts of progress that are being made in undertaking these forms of analysis.

In conclusion, the economists involved in producing this volume have attempted to bring the most advanced and appropriate tools to bear on the issue of resource allocation involving a complex commodity. We hope that these papers demonstrate some of the ways to think about this complexity, and to deal with the important economic issues emanating from that complexity. There remains to be accomplished large amounts of substantive research on the topic of efficient water resource management. We hope that these papers suffice to generate a veritable flood of high quality research on this important topic.

Notes

[1] In this respect, some papers help bring forward the lack of appropriate data, in terms of both quantity and quality, as a major problem in the use of quantitative economic techniques for the analysis of water scarcity problems.

[2] These authors apply a two-stage model in which first the choice of price block is estimated using Probit analysis and then the quantity within that block is estimated by including Mill's ratio among the explanatory variables to account for endogeneity.

[3] Exceptions include Agthe and Billings (1980) and Carver and Boland (1980) who estimate dynamic demand for water, albeit based on intertemporally 'naïve' specifications of consumer behaviour based on Koyck, flow adjustment and stock adjustment assumptions.

[4] It has been found that, for surface irrigation methods, average irrigation water efficiency is about 0.6, whereas drip or sprinkler technologies may increase efficiency up to 0.95.

[5] See Basar and Olsder (1982) for definitions.

References

Agthe, D.E. and Billings, R.B. (1980) Dynamic models of residential water demand, *Water Resources Research* **16**(3), 476–480.

Basar, T. and Olsder, G.J. (1982) *Dynamic Non-Cooperative Game Theory*, Academic Press, New York.

Battese, G.E. and Coelli, T.J. (1995) A model for technical inefficiency effects in a stochastic frontier production function for panel data, *Empirical Economics* **20**, 325–332.

Billings, B. (1982) Specification of block rate price variables in demand models, *Land Economics* **58**, 386–393.

Bockstael, N.E., Hanemann, W.M. and Stand, I.E. (1987) Measuring the benefits of water quality improvements using recreation demand models, Environmental Protection Agency Cooperative Agreement CR-811043-01-0.

Carver, J.H. and Boland, J.J. (1980) Short- and long-run effects of price on municipal water use, *Water Resources Research* **16**(4), 609–616.

Caswell M.F. and Zilberman, D. (1986) The effects of well depth and land quality on the choice of irrigation technology, *American Journal of Agricultural Economics* **68**, 798–811.

Caulkins P.P., Bishop, R.C. and Bouwes, N.W. Sr. (1986) The travel cost model for lake recreation: A comparison of two methods for incorporating site quality and substitution effects, *American Journal of Agricultural Economics* **68**, 291–297.

Coral, L., Fisher, A. and Hatch, N. (1998) Price and non-price influences on water conservation: An econometric model of aggregate demand under nonlinear budget constraint, Working Paper No. 881, Department of Agricultural and Resource Economics and Policy, University of California at Berkeley.

Debreu, G. (1951) The coefficient of resource utilization, *Econometrica* **19**, 273–292.

Dockner, E. and Van Long, N. (1993) International pollution control: Cooperative versus noncooperative strategies, *Journal of Environmental Economics and Management* **25**, 13–29.

Farrell, M.J. (1957) The measurement of productive efficiency, *The Journal of the Royal Statistical Society, Series A, General* **120**(3) 253–281.

Feinermann, E. and Knapp, K.C. (1983) Benefits from groundwater management: Magnitude, sensitivity and distribution, *American Journal of Agricultural Economics* **65**, 703–710.

Griffin, R. and Bromley, D. (1982) Agricultural runoff as a nonpoint externality: A theoretical development, *American Journal of Agricultural Economics* **64**, 547–552.

Hassine, N. and Thomas, A. (1997) Agricultural production, attitude towards risk, and the demand for irrigation water: The case of Tunisia, Working paper, University of Toulouse.

Hausman, J. (1978) Specification tests in econometrics, *Econometrica* **46**, 1251–1271.

Hausman, J., Leonard, G. and McFadden, D. (1992) A utility consistent, combined discrete choice and count data model: Assessing recreational use losses due to natural resource damage, Paper presented at the Cambridge Economics Symposium titled "Contigent Valuation: A Critical Assessment", Washington, DC.

Herrington, P.R. (1987) *Pricing of Water Services*, OECD, Paris.

Hewitt, J. and Hanemann, W. (1995) A discrete/continuous choice approach to residential water demand under block rate pricing, *Land Economics* **71**(2), 173–192.

Howitt, R.E., Watson, W.D. and Adams, R.M. (1980) A reevaluation of price elasticities for irrigation water, *Water Resources Research* **16**, 623–628.

King, D.A. and Sinden, J.A. (1988) Influence of soil conservation on farm land values, *Land Economics* **64**, 242–255.

Koopmans, T.C. (1951) An analysis of production as an efficient combination of activities, in T.C. Koopmans (ed.), *Activity Analysis of Production and Allocation*, Cowles Commission for Research in Economics, Monograph No. 13, Wiley, New York.

Kopp, R.J. (1981) The measurement of productive efficiency: A reconsideration,Ť Quarterly Journal of Economics, 96: 477-503

Martinez-Espineira, R. (2000) Residential water demand in the Northwest of Spain, Paper presented in the 10th Annual Conference of the European Association of Environmental and Resource Economists. Also appears in this volume.

Moore, M.R. and Negri, D.H. (1992) A multicrop production model of irrigated agriculture, applied to water allocation policy of the bureau of reclamation, *Journal of Agricultural and Resource Economics* **17**, 30–43.

Moore, M.R., Gollehon, N.R. and Carrey, M.B. (1994) Alternative models of input allocation in multicrop systems: Irrigation water in the Central Plains, United States, *American Journal of Agricultural Economics* **11**, 143–158.

Muellbauer, J. and Pashardes, P. (1992) Tests of dynamic specification and homogeneity in demand systems, in L. Phlips and L.D. Taylor (eds.), *Aggregation, Consumption and Trade: Essays in*

Honour of Hendrik Houthakker, Advanced Studies in Theoretical and Applied Econometrics, Kluwer Academic Publishers, Dordrecht, pp. 55–98.

Negri, D.H. (1989) The common property aquifer as a differential game, *Water Resources Research* **25**, 25–29.

Neher, P. (1990) *Natural Resource Economics, Conservation and Exploitation*, Cambridge University Press, Cambridge.

Nordin, J. (1976) A proposed modification on Taylor's demand supply analysis: Comment, *Bell Journal of Economic Management and Science* 7(2), 719–721.

Pint, E.M. (1999) Household responses to increased water rates during the California drought, *Land Economics* **75**, 246–266.

Provencher, B. (1995) Issues in the conjunctive use of surface water and groundwater, in D. Bromley (ed.), *The Handbook of Environmental Economics*, Blackwell, Oxford.

Provencher, B. and Burt, O. (1993) The externalities associated with the common property exploitation of groundwater, *Journal of Environmental Economics and Management* **24**, 139–158.

Schefter, J. and David, E. (1985) Estimating residential water demand under multi-tariffs using aggregate data, *Land Economics* **61**, 272–280.

Shunway, C.R. (1973) Derived demand for irrigation water: The California aqueduct, *Southwestern Journal of Agricultural Economics* **5**, 195–200.

Smith, V.K. and Desvousges, W.H. (1986) *Measuring Water Quality Benefits*, Kluwer Academic Publishers, Boston, MA.

Swanson, T. and Vighi, M. (1999) *Regulating Chemical Accumulation in the Environment*, Cambridge University Press, Cambridge.

Taylor, L.D. (1975) The demand elasticity: A survey, *The Bell Journal of Economics* **6**, 74–110.

Thomas, A. (1995) Regulating pollution under asymmetric information: The case of industrial wastewater treatment, *Journal of Environmental Economics and Management* **28**, 357–373.

Torell A., Libbin, J. and Miller, M. (1990) The market value of water in the Ogallala Acquifer, *Land Economics* **66**, 163–175.

Tsutsui, S. and Mino, K. (1990) Nonlinear strategies in dynamic duopolistic competition with sticky prices, *Journal of Economic Theory* **52**, 131–161.

Xepapadeas, A. (1992) Environmental policy design and dynamic nonpoint source pollution, *Journal of Environmental Economics and Management*.

Young, R.A. and Haveman, R. (1985) Economics of water resources: A survey, in A.V. Kneese and J. Sweeney (eds.), *Handbook of Natural Resource and Energy Economics, Volume II*, North Holland, Amsterdam.

Zilberman, D., Shah, F. and Chakravorty, U. (1997) Efficient management of water in agriculture, in D.D. Parker and Y. Tsur (eds.), *Decentralization and Coordination of Water Resource Management*, Kluwer Academic Publishers, Dordrecht, pp. 221–246.

PART 2: DEMAND FOR WATER

2.1. *Residential Demand for Water*

Estimating Water Demand under Increasing-Block Tariffs Using Aggregate Data and Proportions of Users per Block *

R. MARTÍNEZ-ESPIÑEIRA

1. Introduction

Forming a strategy for the management of a water resource requires reasonable knowledge of the effects that alterations in the control variables bring on water use. Such knowledge of the determinants influencing domestic water demand constitutes an essential ingredient of a pricing policy for domestic water supply. Special attention should be paid to the effects of the level, shape, and frequency of the tariff, since these are the variables that can be altered by the policy-maker.

Many studies have dealt with the estimation of residential water demand functions. Most of them, however, have used USA data. Empirical evidence on European countries is very scarce (see OECD, 1999, for a compilation of studies). In this paper, the domestic water demand function in three towns of the Northwest of Spain is estimated.

There are two innovative features in this empirical application that constitute a relevant contribution in the context of European studies of water demand. Both features are expected to result in a more theoretically sound estimation of water demand, at the expense of availability of information not available to the researcher in previous studies.

The first one, the main objective of the paper, is the analysis of the marginal price elasticity and the relation between the coefficients of income and the difference variable when a theoretically correct specification of the price variables for studies on aggregate data under block tariffs is used. This specification, suggested by Schefter and David (1985), includes a weighted-mean marginal price and a weighted-mean difference (as defined by Nordin, 1976). It uses as weights the proportion of users falling in each block of consumption, which is a very unusual type of information. Only one other study (Corral et al., 1998), based on data from the United States, has used this specification with real data. To the author's

* ESRC (Economic and Social Research Council) and the Ilma. Diputación de A Coruña have funded parts of this research. This has benefited from supervision, suggestions, unpublished material, comments etc. of the following individuals: Jack Pezzey, Charles Perrings, Keith Hartley, Lena Höglund, Céline Nauges, Panos Pashardes and Nile Hatch. All remaining errors and views expressed remain the sole responsibility of the author.

P. Pashardes et al. (eds.),
Current Issues in the Economics of Water Resource Management, 27–46.

knowledge, this is the first estimation of a water demand function under non linear tariffs in Europe that uses the theoretically correct price specification suggested in the literature.

The results obtained with the innovative correct specification are compared with those obtained using, for the same data set, a more conventional price specification. This comparison confirms that, as many other studies have predicted, models based on the assumption of a perfectly informed consumer, although theoretically sound, could be less useful than other models less informationally demanding that assume an imperfect knowledge of the tariff.

The second contribution is due to the fact that the estimations of the water demand functions are based on quarterly billing data transformed into their monthly equivalents. The use of monthly (instead of annual) data constitutes an important innovation within the European context. Previous studies based on an aggregate approach have only employed annual data, while the only case study that used biannual data (Hanke and de Maré, 1982) used household-level data for a single town. However, it is very important to analyse intra-annual data in order to approximate more appropriately the seasonal dimension of water demand.

2. Background Literature

One of the most controversial issues faced by the literature on water demand is the analysis of block tariffs. The problems surrounding the theoretical modelling of demands under this tariff type and the complexities that arise in empirical studies based on real data are the subject of a great deal of studies.

The earliest studies normally use the marginal price or average price corresponding to the block where the user (or, with aggregate data, the typical user) consumes. However, following Taylor (1975) and Nordin (1976), a *difference* variable is now commonly added to account for the effect of intramarginal rates. Specific for the case of multiple-block tariffs (and those cases where a fixed quota and/or a free allowance is used), the difference variable is the difference between the total bill and what the user would have to pay if all units were charged at the marginal price. This variable should represent the income effect imposed by the tariff structure. For increasing-block tariffs (as long as the effect of a fixed quota and or a free allowance does not counteract the effect of the blocks) the difference is negative. It can be seen as a subsidy applied to the first units of consumption in compensation for the higher prices paid within higher blocks. When the tariff is decreasing, the difference is positive and acts like a tax applied to the first units of water used in compensation for the lower prices in the higher blocks. The difference variable takes the value of zero under perfectly linear tariffs.

There has been a debate (e.g. Billings and Agthe, 1980; Foster and Beattie, 1981; Houston, 1982, 1983; Billings, 1983) as to whether it is advisable to introduce this type of variable rather than other alternatives. However, Nordin's approach seems to have reached consensus as the correct measure for the intramarginal variable.

In linear models, the coefficient of the difference variable should be equal in magnitude and opposite in sign to the one of the income variable. This has been tested in the empirical literature, but with little success (Billings and Agthe, 1980; Howe, 1982; Jones and Morris, 1984; Nieswiadomy and Molina, 1988, 1989). This finding has led to different types of reactions. First, lack of consumer information about the tariff structure and the fact that the difference variable amounts to a small fraction of the household income have been suggested as explanations for this (Nieswiadomy and Molina, 1989). Others argue that the effect of the intramarginal price changes is not estimated correctly, and that introduces bias in the results.

The use of aggregate data has also been blamed as a major source of incorrect specification. Schefter and David (1985) argue that the theoretically correct way to introduce Nordin's specification is not to include the marginal price for the typical user and the difference for the typical user. They show that a weighted-mean marginal price and a weighted-mean difference are more appropriate. To calculate the weights for these average measures the researcher needs to have access to information on the number of users consuming in each block and this is normally not available.

Schefter and David (1985) resorted to simulated data on the distribution of water users across blocks to empirically test Nordin's specification. They showed how realistic assumptions on the distribution of users among blocks led to the estimated coefficients of income and difference being equal in size and opposite in sign, as predicted by the theory. In contrast, under the commonly adopted assumption that all users fall in the same block as the typical user, the results for income and difference did not match the theoretical expectations.

Consumer behaviour under multiple-block pricing has been modelled by Burtless and Hausman (1978) and Moffitt (1986, 1990), who propose a two-stage model in which the consumers first select the block, and then maximise their utility subject to a budget constraint. Hewitt and Hanemann (1995) are among the very few who used this kind of model in water demand studies and they applied it to individual data. The first stage of this procedure involves the use of some type of discrete-choice model (Probit, Logit) to estimate the probability of falling in a certain block of consumption.

Schefter and David (1985) made assumptions about the distribution of users among blocks but they did not explicitly model the discrete-choice problem by explaining how the probabilities of falling in a certain block were determined.

Corral et al. (1998) incorporated both improvements by using for the first time real proportions of users per block to weight the average marginal price and difference. They also modelled explicitly the choice of block of consumption. However, they did not try to test the hypothesis of equality of the coefficients associated with the income and difference variables.

3. Methodology

The analysis represents a modification of the one undertaken by Corral et al. (1998), who are the only authors known to have used the proper weighted-mean marginal price with real data. They based their model on a water demand function under non-linear budget constraints inspired by the work of Moffitt (1986) and Hewitt and Hanemann (1995). The main difference between the specification used in Corral et al. (1998) and the one used here is the definition of the difference variable. In our case, the definition is precisely the one proposed by Nordin (1976) and used by Schefter and David (1985): the difference between the actual bill and what the bill would have been had all units been charged at the marginal price. Additionally, the difference variable is kept separate from the income variable as an autonomous argument in the demand function. This keeps the analysis more within the context of the original Nordin–Schefter–David formulation and makes it possible to check whether the coefficients of the income variable and of the difference variable are different or not.

The demand function is determined by the fact that the consumer faces a non-linear budget constraint with m piecewise linear segments:[1]

$$I - d_1 = P_1x + y \text{ if } x \text{ falls within the 1st segment,} \tag{1.1}$$

$$I - d_2 = P_2x + y \text{ if } x \text{ falls within the 2nd segment,} \tag{1.2}$$

$$\dots$$

$$I - d_m = P_mx + y \text{ if } x \text{ falls within the } m\text{th segment,} \tag{1.m}$$

where I is income, P_i are the marginal prices in each block i, x is the amount of water demanded, y is a vector of all goods apart from water, and d_i represents the difference in the ith block as defined by Schefter and David (1985). It is, then, the difference between the actual bill and the product of the amount of water used and the marginal price.

The terms on the left-hand side represent the income minus the difference variable, so they are equivalent to a *virtual income*. This can be allocated to the consumption of water, priced at P_i, and all other goods. The virtual income and the marginal price of water vary according to the block considered.

There is another difference between the definition of difference used here and definitions used elsewhere. Some tariff structures in the sample use a minimum of consumption (effectively equivalent to a fixed quota plus an initial free allowance).

Therefore, there is no straightforward way to calculate the value of difference for users falling within the minimum. One option is to assume that the users recognize the effective marginal price of water within the minimum as null (and react accordingly), but this is unrealistic. On the other hand, assuming that the marginal price to which the *within-minimum* users react is the ratio of the total minimum quota to the number of cubic metres in the free allowance amounts to admitting that the difference does vary within the minimum according to the amount of water actually used.

To avoid this difficulty, the value of difference for the typical *within-minimum* user is used as the value of difference for the blocks associated with the minimum. This should theoretically be calculated as the ratio of the number of cubic meters used within the first block (or minimum) by the *within-minimum* users to the number of these users. However, this definition led to distorted results, probably because it overemphasizes the effect of those users who do not use any water at all[2] leading to an overestimation of the difference for users of positive amounts of water.

The compromise solution adopted is the use of the arithmetical unweighted mean of the amount of water included in the minimum. It is assumed that the typical user of positive amounts of water that ends up in the minimum uses this amount of water. This is, then, a measure that varies only with the change of the tariff structure, so the component of the difference stemming from the users in the minimum varies only when the price is updated. This affects only the users in the minimum and only for half of the observations, since the other half do not have minima in use.

Note again that, under conditions of perfect information, the consumer would recognize that the price within the minimum is actually zero at the margin and the real difference would be constant at the level of the minimum bill (including some extra fixed quotas if any). The results of the estimation conducted under this assumption are in any case very similar to the ones here reported (the issue affects only some observations), although the marginal price estimate is somewhat less significant. These results are available upon request.

Given the nondifferentiability of the budget constraint, the optimization process is assumed to include two stages (Moffitt, 1986, 1990; Hewitt and Hanemann, 1995; Corral et al., 1998). First, the consumers choose in a continuous way the particular level of use that is optimal within each segment of the kinked budget constraint. This first stage results in the conditional demand function. In a second stage, the consumer discretely chooses the conditional demand that maximizes overall utility.

This function can be expressed as:

$$x = b_1 x_1^*(P_1, I - d_1) + \ldots b_m x_m^*(P_m I - d_m) + c_1 X_1 + \cdots + c_{m-1} X_{m-1}, \quad (2)$$

where x_i^* gives the optimal level of consumption conditional on being located within the ith segment (where $i = 1, 2, \ldots, m$) and X_i denotes the upper limit of each segment.

Also:

$$b_1 = 1 \text{ if } x_1^*(P_1, I - d_1) < X_1 \text{ and } b_1 = 0 \text{ otherwise,} \tag{3a}$$

$$b_i = 1 \text{ if } B_{i1}, B_{i2} > 0 \text{ and } b_i = 0 \text{ otherwise,} \quad (i = 2, 3, \ldots, m - 1), \tag{3b}$$

$$b_m = 1 \text{ if } X_{m-1} < x_m^*(P_m, I - d_m) \text{ and } b_m = 0 \text{ otherwise,} \tag{3c}$$

$$c_i = 1 \text{ if } C_{i1}, C_{i2} > 0 \text{ and } c_i = 0 \text{ otherwise (for } i = 1, 2, \ldots, m), \tag{3d}$$

$$B_{i1} = X_i - x_i^*(P_i, I - d_i), \tag{3e}$$

$$B_{i2} = x_i^*(P_i, I - d_i) - X_{i-1}, \tag{3f}$$

$$C_{i1} = x_i^*(P_i, I - d_i) - X_i, \tag{3g}$$

$$C_{i2} = X_i - x_{i+1}^*(P_{i+1}, I - d_{i+1}). \tag{3h}$$

The data employed are aggregate at the municipal level. Therefore, to derive an econometrically estimable model from it, this function must be transformed by summing over the number of users in each municipality in each period. Let the total number of users in each community (measured by the variable *ABONS*) be N. The aggregate demand function, based on increasing-block tariffs (the ones used in the sample) and yielding a convex budget set, becomes then:

$$X = \sum_{j=1}^{n} b_1 x_{j1}^*(P_1, I - d_1) + \cdots + b_m x_{jm}^*(P_m, I - d_m) \tag{4a}$$

$$= Q_1(P_1, I - d_1) + \cdots + Q_m(P_m, I - d_m) \tag{4b}$$

$$= n_1 q_1(P_1, I - d_1) + \cdots + n_m q_m(P_m, I - d_m), \tag{4c}$$

where $x_{ji}^*(\cdot)$ is the conditional demand of the jth consumer (where $j = 1, \ldots, N$) in the ith block, $Q_i = \sum_{j=1}^{n} b_i x_{ji}^*(\cdot)$ and n_i and q_i are the number of users and the average use in the ith segment. The discrete-choice part of the user's decision is assumed to determine the number of users that end up consuming on the ith segment n_i. The continuous-choice part of the problem defines the average consumption $q_i(\cdot)$ conditional on being located in the ith block. The structure of the individual unconditional demand is preserved, although the consumers located at the kinks X_i cannot be considered. This is because the data, due to their aggregate nature, lack information on those consumers. Corral et al. (1998) acknowledge the same problem.

A normalisation to account for the different number of users in different towns and periods is required. This yields the following aggregate demand function:

$$q = s_1 q_1(P_1, I - d_1) + s_2 q_2(P_2, I - d_2) + \cdots + s_m q_m(P_m, I - d_m), \quad (5)$$

where q is the average use per account and $s_i = n_i/N$ is the proportion of consumers who end in the ith block.

This can be transformed into:

$$q = s_{1t} q_{1t}(P_{1t}, I - d_{1t}, Z_t|\beta) + \cdots + s_{mt} q_{mt}(P_{mt}, I - d_{mt}, Z_t|\beta) + \epsilon_t, \quad (6)$$

where t is the time index, Z represents a matrix of climate and sociodemographic variables, β is the vector of coefficients to be estimated and ϵ_t is the error term.

If a further assumption of linearity of the demand function is made, the form of the function simplifies to:

$$qt = \beta_0 + \beta_1 \sum_{i=1}^{m} s_{it} p_{it} + \beta_2 \sum_{i=1}^{m} s_{it}(I - d_{it}) + \gamma Z_t + \epsilon_t, \quad (7)$$

where γ is a vector of parameters associated with Z.

A further transformation isolates income from difference so that they keep independent coefficients:

$$qt = \beta_0 + \beta_1 \sum_{i=1}^{m} s_{it} p_{it} + \beta_2 \overline{I} + \beta_3 \sum_{i=1}^{m} s_{it} d_{it} + \gamma Z_t + \epsilon_t, \quad (8)$$

where $\overline{I}$ is simply the average income in the community. Note that, theoretically, $\beta_3 = -\beta_2$.

It would not be correct to estimate this equation using s_{it}, the observed proportions or probabilities of ending in each block of consumption, since they are, like the conditional demands, functions of preferences and therefore correlated with the error term. The equation should itself be estimated in two stages. First the proportion of users ending in each block should be estimated. Then the predicted proportions for each town and each period stemming from the first step can be used to estimate the unconditional demand function for all the towns. The first step is undertaken by applying a Multinomial Logit Model with, in this case, grouped data of the type:

$$\Pr(\text{choice}) = \frac{e^{\beta_i' x_t}}{\sum_i e^{\beta_i' x_t}}, \quad i = 1, 2, \ldots, M, \quad (9)$$

in which the observed dependent variable is the proportion of consumers ending in each block and X includes month-specific climatic variables and town-specific sociodemographic variables.

Once the first step of the estimation described above is completed, the predicted proportions of users per block are estimated and used to calculate the weighted-mean marginal price and weighted-mean difference, as proposed by Schefter and David (1985).

In a second step the demand function is estimated using panel-data analysis. The price variables calculated in the first step are introduced alongside climate, demographic and billing variables. The estimation techniques employed take into consideration the presence of non-spherical disturbances in the model (see Greene, 1991, 1993).[3]

4. Data

Confidential data on consumption at the municipal level are combined with tariffs to obtain the billing variables. Sociodemographic information at the municipal level and monthly data on climatic variables stemming from the closest meteorological station was also collected.

Only three municipalities were able to provide appropriate data on the proportion of users per block. The users from one of them could be, however, split into two groups because some users did not have sewage services.[4] The effective number of units is therefore four, with variable number of months with information available. The data refer to the period January 1995–June 1999 for the community with the greatest number of data points. The total number of observations in the unbalanced panel is 183. This data set is used in the Multinomial Logit. However, in order to conduct the complex panel-data analysis of the second step, the panel had to be balanced into a smaller set of 120 observations (30 observations per community). The smaller data set includes data from March 1996 to September 1998.

The data on sociodemographic variables come from the 1996 census, so there is no within-unit variability for sociodemographic variables. In all cases an increasing-block tariff is used and only two of the four communities have no minimum of consumption. A description on the definitions of the variables used and a summary of their values can be found in the Appendix.

5. Estimation and Results

In the Multinomial Logit (see Appendix for a summary of results) 183 observations are used to predict the proportions (*PROABi*) used as weights for the average marginal prices and differences applying to each block, using as explanatory variables *INCOME*, *UND*19, *OV*64, and *AVTEMP*. Note that no price variables act as regressors, since they are found to be highly non-significant and/or non-conclusive when introduced. This agrees with other results found in the literature (see, for example,

Houston, 1982, in the context of demand for electricity) that assume or prove that it is not the price that really determines the consumption block in which users end up consuming. Also the very usual direct applications of Nordin's approach to demand estimation assume that there are no shifts in the distribution of consumers as a result of price changes. However, it is important to stress that the effect of prices on block distribution could be different if more (and smaller) blocks were included in the tariff.

As expected, the coefficients of *INCOME*, *UND*19, and *AVTEMP* have a positive effect on the proportion of users moving into the higher blocks and a negative effect on the proportion of users remaining in the first block. The coefficient of *AVTEMP* is not found significantly different from zero. All the other coefficients, except for the one of *UND*19, are found significant when analysing the effects on the use in the uppermost block. The final predictions obtained from the Multinomial Logit are reasonably close to the originally observed values. The introduction of additional variables proved unadvisable, given the likely strong levels of multicolinearity between different variables, the limited number of observations, and the lack of within-town variability of the sociodemographic variables.

The second step involves the calculation of *WEMAP* and *WEDIFF* using the predicted proportions and the variables on prices and differences in each block. The initial variables are deflated into constant 1992 prices. These price variables correspond to the theoretical specification proposed by Nordin (1976) to estimate demand in the presence of block tariffs using aggregate data, as Houston (1982) and Schefter and David (1985) first pointed out.

The aggregate demand function is estimated with *MONTHLY* as the dependent variable and *WEMAP* and *WEDIFF* (both price variables), *OV*64, *INCOME*, and *AVTEMP* as the independent variables. The panel is analysed using groupwise regression models that account for heteroskedasticity and cross-sectional correlation, also testing the potential presence of autocorrelation (with one, two and three lags). The results of the Breusch–Godfrey test (Breusch, 1978; Godfrey, 1978) reveal that only the autocorrelation with one lag is highly significant.[5] This is obviously due to the way in which quarterly observations of water use are transformed into three monthly (equal) equivalents. Different combinations of assumptions about the matrix of disturbances were checked (see Greene, 1991). The final model assumes a groupwise heteroskedastic and cross-sectionally correlated model with first-order autocorrelation, whose degree is assumed to be specific for all four towns. After checking nine different combinations of the specifications of the disturbances matrix, the preceding one is chosen both because of the plausible results and because it presents the second highest value of the log-likelihood estimate for all price specifications. The log-likelihood value would be slightly higher under the assumption of common degrees of autocorrelation (not a very realistic assumption that yields rather poor results).

The specification finally selected (to which the results summarized below refer) considers first-order group-specific autocorrelation, heteroskedasticity and cross-sectional correlation. However, a pooled OLS estimation and a two-step GLS groupwise regression that assumes no autocorrelation and homoskedasticity are also presented, for comparison, in the Appendix, where detailed results are reported.

The coefficients of *AVTEMP*, *OV64*, and *INCOME* have the expected signs. The higher the monthly average temperature and the average level of income, the more the consumption. Greater proportions of retired people in the community seem to decrease average water use. However, only the estimate on *AVTEMP* is highly significant (at the 1% level), while (at the 10% level) it cannot be rejected that the effect of *INCOME* and *OV64* is null. The estimated coefficient of *WEDIFF* is highly non-significant and presents a wrong sign. A comparison between the magnitude of the coefficient of *INCOME* (once transformed into equivalents per month per capita[6]) and the coefficient for *WEDIFF* is attempted, but their magnitudes (as the tables show) are very different. This result once again casts doubts on the specification originally proposed by Nordin (1976). However, normally the size of the estimate of *INCOME* should be expected to appear greater than the size of the one of *WEDIFF*. Surprisingly, this expectation is not matched by the results for any of the price specifications compared.[7] This could be because of the lack of variation for *INCOME* available in the sample.

However, attention is focused on the coefficient of the marginal price variable *WEMAP*. This has the correct sign and a reasonable magnitude in the three models. Its size becomes more realistic when the residuals are treated with more accuracy. In the finally chosen model, the hypothesis that the coefficient of *WEMAP* is actually zero can be rejected with more than a 99% level of confidence. This estimate results in an elasticity (calculated at the means of price and use) of −0.47. This is a somewhat high estimate for the marginal price, explained by the fact that the communities in the sample use relatively high levels of prices and increasing tariffs. It should also be considered with caution, since only 120 observations are used.

The estimates for *WEMAP* can be compared with the results found for another, more conventional, price specification. As shown in the Appendix, the price variables associated with the typical use *TYPDIFF* and *TYPMGP* have correct signs and highly significant estimates. In this specification the marginal-price elasticity at the mean price and mean use is −0.37. This estimate is relatively close to the one for *WEMAP* in this particular sample. In fact the 95% confidence interval for this marginal price estimate embraces the values of elasticities within the range −0.54 to −0.21. Also the 95% confidence interval for *WEMAP* is (−0.82, −0.13). Therefore, the hypothesis that the *WE* specification and the more conventional *TYP* one actually yield the same elasticities cannot be rejected.

The elasticity results are reported at the means of use and prices, which is the common procedure applied under linear functional forms. However, it could be

Table 1. Elasticity values for extreme values of prices.

Price specification	95% Confidence interval point	Minimum price	Mean price	Maximum price
WEMAP	high	–0.669	–0.822	–2.204
	central	–0.312	–0.475	–0.659
	low	–0.068	–0.127	–0.119
TYPMGP	high	–0.134	–0.536	–0.322
	central	–0.089	–0.373	–0.204
	low	–0.048	–0.210	–0.105

argued that many times econometric specifications show their relative strength not in the prediction of averages but in the prediction of values in the tails of parameter distributions.[8] For this reason, the 95% confidence intervals of the value of the price elasticities are calculated for the highest and lowest prices found in the sample.

These results (presented in Table 1) show that when the maximum or minimum values of price found in the sample (and their predicted *MONTHLY* values) are used to calculate the elasticities, the hypothesis of equality between the results of both specifications cannot be rejected either.

Under the *TYP* specification, the estimates for *INCOME* and *OV64* present the expected signs and are always significant, but the hypothesis that the effect of *AVTEMP* be null cannot be rejected. This is the price specification most commonly used in the literature when only aggregate data are available, hence the interest in comparing it with the *WE* specification. It has to be stressed that the *TYP* specification does not take into account the potential simultaneity between price and quantity. The price for the typical user has also a tendency to vary with the quantity actually demanded. This is expected to dampen the effect of price on use. This simultaneity has been corrected by the two-step procedure applied with *WE* but not when *TYP* is applied.

These marginal-price elasticities fall within the range that could be reasonably expected in a region where increasing tariffs are used. The more theoretically correct price specification yields a higher elasticity because it accounts for potential simultaneity between price and use and because its variables change to a higher extent from one billing period to the next. The specification *TYP* basically considers only marginal price in the block where the typical user falls (ignoring how much typical use actually is as long as it does not cross the border of the block) and ignores any redistribution of consumers among blocks.

When the instrumental price variables (*IV*) stemming from a linearisation of the tariff structure (proposed by Billings, 1982) is used, the coefficients of these price variables are non-significant and the model performs rather poorly. This can be partly explained by the low degree of variability of the *IV* price variables combined with the low number of observations used. The *IV* variables vary only annually (when the price level changes) and do not depend on the monthly variations of consumption (as the *WE* and *TYP* price variables do). The results obtained with this specification are not reported but are available on request.

The inequality between the sizes of the coefficients of *INCOME* and the difference variable is blatant throughout all specifications. This confirms that the original Nordin (1976) specification, although theoretically sound, could be not the best under conditions of imperfect information about the tariff.

6. Conclusions

In this paper, the price specification suggested as theoretically correct when analysing demands under block tariffs with aggregate data is applied to a small sample of monthly observations. This enables marginal price and difference to be introduced in a more appropriate manner than in previous empirical studies.

The hypothesis of equality between the absolute values of the coefficients of income and difference cannot be rejected. This result agrees with previous studies, supporting the idea that the specification suggested by Nordin (1976) might not be applicable to the case of water demand, due to the lack of information on the tariff shape available to the user. In this particular case, Schefter and David's (1985) explanation for the mismatches between theoretical expectations and empirical results, based on problems of misspecification of the price variables, does not seem to apply.

The results obtained with the most innovative price specification are compared with the results obtained with a more conventional one. These comparisons show that the estimated elasticities may not differ significantly when a more conventional price specification is used. They also show that simpler and less data-intensive models can better explain the variation in water demand.

The estimated influence of climate and sociodemographic variables confirm *a priori* expectations. Higher proportions of youngsters, higher temperatures, and higher average income tend to increase the proportion of users falling in higher use-blocks. Higher proportions of retired people are associated with a lower proportion of users in the higher use-blocks. These variables exert analogous effects on average monthly use.

The marginal-price elasticities estimated fall within the range that could reasonably be expected in a region where increasing tariffs and high price levels are used. The more theoretically correct price specification yields a higher elasticity because it accounts for potential simultaneity between price and use and because its variables change to a higher extent between billing periods. However, the hypothesis that the elasticities found using the different specifications are actually equal to each other cannot be rejected.[9]

It would be advisable to undertake the same kind of comparison using non-linear demand functions, since it is possible that the problem of aggregation is not so acute for a linear shape. This study is restricted to the linear functional from for the sake of comparability with the majority of previous studies. It would have been desirable to use another very common functional form, based on logarithmic transformations, but the presence of positive and negative values in the difference variables rules it out.

This is the first time that the theoretically correct price specification for demand functions under block pricing and aggregate data has been applied to European data. It would be desirable to conduct further studies applying the same methodology to different geographical areas so that results can be compared. In particular, tests should be applied to samples facing tariffs with more and smaller blocks as well as both higher and lower price levels. Also alternative econometric specifications (perhaps based on the availability of a larger sample) could be applied. This would help to determine the generality of the results reported here.

Some policy implications for water management can be derived. One is that water management should still rely on research based on data on the proportion of users per block, when available. However, the results above suggest that when these data are only partially available, it might be better to resort to more conventional price-specifications. This is because, in many instances, the level of prices, the difference between block-prices, the width of blocks, and the concentration of users in certain blocks might be such that the bias introduced by a less theoretically sound specification would be relatively small. This bias might be small enough relative to the gain derived from using additional time-series or cross-sectional observations to justify their use.

It could well be the case that data on the distribution of users among blocks were only available for a limited range of a time series or for certain communities in a sample. The costs of retrieving this information from individual bills, if available, would normally be nontrivial. For this reason, a pilot-study on the best-quality sample could be used to estimate whether the bias associated with the use of a less informationally-demanding specification would merit further data collection or processing or, as in the case analysed here, it would be better to also use the data points with no information available on users distribution.

These considerations are relevant because of the high costs associated with the use of household-level data relative to community-level data to analyse water demand.

The results in this paper also suggest that water managers who use block tariffs with only a few wide blocks should consider that changes in prices and block sizes might have little effect of the distribution of users among blocks. This could have important practical consequences when it comes to the modification of these tariffs.

because factors beyond the managers control could exert stronger effects on the distribution of consumers among blocks than changes in price differentials and small changes in block sizes.

7. Appendix

7.1. Variables Definition[10]

ABONS[11] = total number of domestic accounts read in each period.

AVTEMP = average temperature in each month. Source: *Centros Meteorológicos Territoriales* of the *Instituto Nacional de Meteorología* and TEMPUS database of the *INE (Instituto Nacional de Estadística)*. Unit = Degree Celsius/10.

FIXEDP = fixed component of water+sewage bill. This variable does not include the amount to be paid for the units of consumption charged within the minimum (although a minimum amounts effectively to a fixed quota plus a free allowance). The main difference between the fixed quota and the effect of the minimum is that the fixed quota keeps on playing its role for the whole range of consumption while the fixed quota effect of the minimum disappears when the user goes beyond the minimum threshold. VAT is added and the provincial variations of the RPI are used to deflate values. Source: see note 11. Unit = 1992 pesetas (ESP).

INCOME = index for estimated family disposable income per capita in the municipality. It consists of a set of intervals of family disposable income per capita within which the municipalities are assumed to fall. The data contain only two values of *INCOME* so they were transformed into the medium value of the associated interval. These are 4 = 81187 ESP (1992)/month and 7 = 116226 ESP (1992)/month. Source: La Caixa (1999).

M3CAPITA = per account water registered as used in the community in a particular billing period. It is constructed as the total of registered domestic use divided by *ABONS*. Source: see note 11. Unit = m^3/billing period (which is variable).

MONTHLY = constructed as *M3CAPITA* × (*FRQBILL*/12). It is the dependent variable in the majority of estimations. Source: see note 11. Unit = m^3/month.

OV64 = percentage of population over 64 at 1-5-96. Unit = %. Source: IGE (1998) and EUSTAT (1998).

PBLN = (water+sewage) price of the units sold within each block *N* (block 1 sometimes coincides with the minimum). VAT is added and the provincial variations of the RPI (INE, 1999) are used to deflate values. Source: see note 11. Unit = 1992 ESP/m^3.

PROABi = Proportion of users located in the *i*th block of consumption.

SIZEMIN = number of units to pay for irrespective of actual use. Source: see note 11. Unit = m^3.

TYPDIFF = *Nordin-difference* faced by typical user. It is the difference between what the typical users pay minus what they would have to pay had all units been charged at *TYPMGP*. Unit = 1992 ESP.

TYPMGP = marginal price associated with MONTHLY use, that is, the marginal price faced by the *average* or *typical* consumer. When the average users do not reach the upper limit of the minimum block, it is assumed that they face a marginal price equal to the result of dividing the minimum bill minus *FIXEDP* between the number of units in the minimum. With perfect information, the effective marginal price within the minimum would be zero. Unit = 1992 ESP/m^3.

*UND*19 = percentage of population under the age of 19 (15 in the case the 108 out of 183 observations stemming from Galicia) at 1-5-96. Unit = %. Source: IGE (1998) and EUSTAT (1998).

WEDIFF = mean difference for water and sewage obtained by weighting the difference in each block by the proportion of users falling in that block. In the case of the blocks corresponding to a minimum, the arithmetical unweighted mean of the amount of water included in the minimum is used to calculate the difference associated with that block.[12] Unit = 1992 ESP.

WEMAP = mean marginal price obtained by weighting the price in each block by the proportion of users falling in that block. Unit = 1992 ESP/m^3.

Table 2. Summary of variables.

Variable	Mean	Std. Dev.	Minimum	Maximum	Cases
AVTEMP	148.183333	33.4178997	80.0000000	209.000000	120
UND19	0.19431750	0.01087343	0.18374000	0.20817000	120
OV64	0.15115000	0.02528179	0.12670000	0.18409000	120
INCOME	98706.6101	17592.8118	81187.2551	116225.965	120
WEMAP	65.3622266	10.3104104	52.3358009	87.4947551	120
WEDIFF	60.1559944	120.090006	-97.2934471	277.777264	120
TYPMGP	65.9787175	12.2918660	45.8100000	94.6600000	120
TYPDIFF	-11.2925833	123.937424	-174.280000	170.560000	120
MONTHLY	10.4583212	2.65701435	4.39089000	19.7592000	120
ABONS	4770.25000	1968.09964	2114.00000	6859.00000	120
PROAB1	0.56868688	0.15022588	0.36705000	0.94280000	183
PROAB2	0.36105065	0.11024030	0.04728000	0.46976000	183
PROAB3	0.00702621	0.06976219	0.00938000	0.24358000	183

Table 3. Multinomial Logit estimation: Results of the block choice model estimation.

Multinomial Logit Model	
Maximum Likelihood Estimates	
Dependent variable	PROAB1
Weighting variable	ABONS
Number of observations	183
Iterations completed	6
Log likelihood function	−149.904
Restricted log likelihood	−160.194
χ^2	20.581
Degrees of freedom	6
Significance level	0.0021811

Variable/Marginal effects	on Prob[block 1]	on Prob[block 2]	on Prob[block 3]
INCOME	−0.3292632	0.2092458	0.1200174
	−4.064***	2.723***	3.789***
UND19	−8.6858385	8.1891762	0.4966623
	−3.137***	2.996***	0.345
OV64	25.3596975	−19.1421466	−6.2175508
	4.406***	−3.413***	−2.466**
AVTEMP	−0.0007769	0.0004626	0.3143846
	−0.688	0.432	0.681

Partial derivatives of probabilities with respect to the vector of characteristics. They are computed at the means of the *Xs*. Observations used for means are All Observations. A full set is given for the entire set of outcomes, $PROAB1 = 0$ to $PROAB1 = 2$. Probabilities at the mean vector are $0 = 0.595$, $1 = 0.338$, $2 = 0.066$.
Values on second rows show t-ratios.
* = significant at 10%, ** = significant at 5%, *** = significant at 1%.

Table 4. Results specification *WE* (with appropriate weighting by the proportions of users per block).

Model Variables	OLS	2-step GLS homoskedastic no autocorrelation	2-step GLS heteroskedastic group correlated autocorrelation
Constant	34.4586521	34.8745607	19.1429712
	4.020***	3.004***	2.373**
	(21.7451143	(20.8588616	(17.2334797
	11.687***)	11.105***)	7.760***)
AVTEMP	0.0291245	0.0353695	0.0146733
	5.095***	4.513***	2.619***
OV64	-298.1079657	-321.8204292	-70.9765432
	-1.924*	-1.516	-0.504
INCOME	4.3078929	4.7491212	0.6470178
	(0.0003688)	(0.0004066)	(0.0000554)
	1.615	1.299	0.268
WEDIFF	-0.0004152	-0.001593	0.0111659
	-0.045	-0.127	1.342
WEMAP	-0.1059656451	-0.1076858	-0.0759566
	-3.530***	-2.839***	-2.708***
N	120	120	120
Log-likelihood	-207.7526	-226.610911	-140.992409
Adjusted R-squared	0.684	NA	NA
Price elasticity	-0.662	-0.673	-0.475

Values on second rows show t-ratios.
* = significant at 10%, ** = significant at 5%, *** = significant at 1%.
Average *MONTHLY* (Dependent) = 10.00026502 (w) 10.4583212 (uw).
Average *WEMAP* = 65.760771 (w) 65.362226 (uw).
w = weighted by *ABONS*; uw = unweighted.
Figures between brackets refer to the estimation when income is estimated using the middle values of the intervals of the index normalised into monthly equivalents. These are for the sample 4 = 81187 ESP (1992)/month and 7 = 116226 ESP (1992)/month.

Table 5. Specification *TYP* (price variables faced by the typical user).

Model Variables	OLS	2-step GLS homoskedastic no autocorrelation	2-step GLS heteroskedastic group correlated autocorrelation
Constant	36.5253988	37.674923	37.343659
	11.364***	11.717***	13.123***
	(16.2492685	(15.6348493	(18.3441056
	8.273***)	8.406***)	12.507***)
AVTEMP	0.0219654	0.0270865	0.0029827
	5.788***	5.767***	1.217
OV64	–412.9787951	–442.9818036	–391.8722106
	–10.578***	–10.720***	–12.060***
INCOME	6.8704243	7.4681242	6.437865201
	(0.0005882)	(0.0006394)	(0.0005512)
	9.566***	9.859***	12.231***
TYPDIFF	–0.0128067	–0.0147622	–0.0108221
	–5.290***	–6.334***	–8.597***
TYPMGP	–0.0737016	–0.0835885	–0.0592014
	–4.308***	–5.076***	–4.535***
N	120	120	120
Log-likelihood	–209.1222	–225.182777	–109.092719
Adjusted R-squared	0.690	NA	NA
Price elasticity	–0.485	–0.527	–0.373

Values on second rows show t-ratios.
* = significant at 10%, ** = significant at 5%, *** = significant at 1%.
Average *MONTHLY* (Dependent) = 10.00026502 (w) 10.4583212 (uw).
Average *TYPMGP* = 65.794479 (w) 65.9787175 (uw).
w = weighted by *ABONS*; uw = unweighted.
Figures between brackets refer to the estimation when income is estimated using the middle values of the intervals of the index normalised into monthly equivalents. These are for the sample 4 = 81187 ESP (1992)/month and 7 = 116226 ESP (1992)/month.

Notes

[1] Note that the terms *segment* and *block* are basically interchangeable, with the exception that the upper bound of the last block is infinite, while the limit of the last segment in the consumers budget constraint, X_m in this paper's notation, is finite.

[2] Because they do not reside in the house for part of the year, for example.

[3] It could be argued that the most appropriate way to estimate the model would be to use a two-error maximum-likelihood technique that simultaneously estimated the discrete-choice and continuous-choice problems (Hewitt and Hanemann, 1995). The two-step procedure described above is used for simplicity and also to keep a greater degree of comparability with Corral et al. (1998).

[4] This group is *less urban* than the one that also paid for sewage. However, no direct account is taken for this, given the lack of sociodemographic data at more than the municipal level. The treatment of heteroskedasticity should suffice to account for the differences in variability of consumption, due mainly to the higher peak consumption in the more rural areas of this community.

[5] The value of the statistic of this test for the *WE* specification is more than 30, clearly higher than the χ^2 critical value of 11.34.

[6] The appropriate measure would be income per household. There would be no exact way to convert the per capita income measure into a per household one. Moreover, that would simply further reduce the value of the coefficient of income around three times, so the exercise would be rather pointless.

[7] Unfortunately, in Corral et al. (1998), authors of the only other study where, to date, the same price specification is used, inserted the difference into the income variable, so no comparison with their results is useful.

[8] I am indebted to Professor P. Pashardes for calling my attention to this issue.

[9] Although these results should be considered with caution, since the size of the sample and the variability of some of the regressors are small.

[10] Further details on sources and variables transformations are available in an extended version of this paper.

[11] The total of domestic registered (and billed) use and the number of water accounts are commercially confidential data provided by the water utilities, consortiums of water suppliers or individual city councils. The prices and tariff structures were provided by the same sources.

[12] This is, then, a measure that varies only with the change of the tariff structure so the component of the difference stemming from the users in the minimum varies only when the price is updated. This affects only the users in the minimum and only for half of the observations, since the other half do not have minima in use.

References

Billings, B. (1982) Specification of block rate price variables in demand models, *Land Economics* **58**(3), 386–393.

Billings, B. (1983) Revenue effects from changes in a declining block pricing structure: Comment, *Land Economics* **59**(3), 351–363.

Billings, B. and Agthe, D. (1980) Price elasticities for water: A case of increasing block rates, *Land Economics* **56**(1), 73–84.

Breusch, T. (1978) Testing for autocorrelation in dynamic linear models, *Australian Economic Papers* **17**, 334–355.

Burtless, G. and Hausman, J. (1978) The effect of taxation on labor supply: Evaluating the Gary negative income tax experiment, *Journal of Political Economy* **86**(6), 1103–1130.

Caixa, La and Instituto Lawrence R. Klein (1999) *Anuario Comercial de España 1999*, Caja de Ahorros y Pensiones de Barcelona, La Caixa, Barcelona.

Corral, L., Fisher, A. and Hatch, N. (1998) Price and non-price influences on water conservation: An econometric model of aggregate demand under nonlinear budget constraint, Working Paper No. 881, Department of Agricultural and Resource Economics and Policy, University of California at Berkeley.

EUSTAT (1998) *Información Municipal*, Instituto Vasco de Estadística (EUSTAT), Euskadi.

Foster, H.S.J. and Beattie, B.R. (1981) On the specification of price in studies of consumer demand under block price scheduling, *Land Economics* **57**(2), 624–629.

Godfrey, L. (1978) Testing against general autoregressive and moving average error models when the regressors include lagged dependent variables, *Econometrica* **46**(6), 1293–1301.

Greene, W.H. (1991) *LIMDEP Version 6.0: User's Manual and Reference Guide*, Econometric Software, Bellport, NY.

Greene, W.H. (1993) *Econometric Analysis*, Prentice Hall, Englewood Cliffs, NJ.

Hanke, S. and de Maré, L. (1982) Residential water demand: A pooled time series cross-section study of Malmö, Sweden, *Water Resources Bulletin* **18**(4), 621–625.

Hewitt, J. and Hanemann, W. (1995) A discrete/continuous choice approach to residential water demand under block rate pricing, *Land Economics* **71**(2), 173–192.

Houston, D. (1982) Revenue effects from changes in a declining block pricing structure, *Land Economics* **58**(3), 351–363.

Houston, D. (1983) Revenue effects from changes in a declining block pricing structure: Reply, *Land Economics* **59**(3), 360–364.

Howe, C. (1982) The impact of price on residential water demand: Some new insights, *Water Resources Research* **18**(4), 713–716.

IGE (1998) Padrón de Habitantes 1996.

INE (1999) IPC, Índices Provinciales. Dato Base (Consumption Price Indices), TEMPUS Database, http://www.ine.es/cgi/menu.pl.

Jones, C. and Morris, J. (1984) Instrumental price estimates and residential water demand, *Water Resources Research* **20**(2), 197–202.

Moffitt, R. (1986) The econometrics of piecewise-linear budget constraints, *Journal of Business and Economic Statistics* **4**(3), 317–328.

Moffitt, R. (1990) The econometrics of kinked budget constraints, *Journal of Economic Perspectives* **4**(2), 119–139.

Nieswiadomy, M. and Molina, D. (1988) Urban water demand estimates under increasing block rates, *Growth Change* **19**(1), 1–12.

Nieswiadomy, M. and Molina, D. (1989) Comparing residential water estimates under decreasing and increasing block rates using household data, *Land Economics* **65**(3), 280–289.

Nordin, J. (1976) A proposed modification on Taylor's demand-supply analysis: Comment, *Bell Journal of Economic Management and Science* **7**(2), 719–721.

OECD (1999) *Household Water Pricing in OECD Countries*, OECD, Paris, Unclassified document ENV/EPOC/GEEI(98)12/FINAL.

Schefter, J. and David, E. (1985) Estimating residential water demand under multi-tariffs using aggregate data, *Land Economics* **61**(3), 272–80.

Taylor, L. (1975) The demand for electricity: A survey, *The Bell Journal of Economics* **6**(1), 74–110.

Long-Run Study of Residential Water Consumption *

CÉLINE NAUGES and ALBAN THOMAS

1. Introduction

Residential water demand has been extensively studied since the end of the sixties. Most of this research has been conducted in the United States (Howe and Linaweaver, 1967; Foster and Beattie, 1979; Chicoine and Ramamurthy, 1986; Nieswiadomy and Molina, 1989; Hewitt and Hanemann, 1995) but studies on European countries have emerged during the nineties only (Point, 1993; Hansen, 1996; Höglund, 1997; CREDOC, 1997; Nauges and Thomas, 2000). The central objective of these papers is the estimation of a residential water demand function where individual or aggregate residential consumption is written as a function of water price and some other variables such as income, household and housing characteristics, domestic appliances and rainfall. Demand function parameter estimates are then used to compute the short-run price elasticity and to assess the impact of demand shifters.

Most of these studies are conducted in a static framework. In other words, current water consumption is specified as a function of current price, income and other socioeconomic variables. However, if current water use is strongly influenced by past water use (through habits or stocks of durables), a dynamic model which explicitly takes this relationship into account may produce superior explanations and predictions of household's behavior. Indeed, it is possible that domestic consumers do not react immediately to variation in prices because residential water use is mainly through durable equipment such as lawns, swimming pools, bathtubs or dishwashing machines. Moreover, the formation of habits could also explain the slow adaptation of behaviour. In many industrialised countries and in France in particular, the price of water has long been very low and domestic consumers have built more or less persistent habits of consumption. The price of water has increased sharply since the beginning of the nineties, after the French government has decided that the price of water should 'reflect its real cost'. This comes as a consequence of the 1992 water act, that called for more efficient water management policies, following a period of severe droughts in the late 1980s. Consumers

* This paper has benefited from comments and suggestions of Alain Carpentier, Alberto Garrido, Roberto Martínez-Espiñeira and an anonymous referee. All remaining errors and views expressed remain the sole responsibility of the authors.

P. Pashardes et al. (eds.),
Current Issues in the Economics of Water Resource Management, 47–66.

have begun to react to price increases but habits are not usually taken off immediately. Agthe and Billings (1980) and Carver and Boland (1980) are among the rare authors to estimate long-run price elasticities using dynamic demand models. Agthe and Billings (1980) test Fisher–Kaysen, Koyck, flow adjustment and stock adjustment econometric models on US data. These authors estimated the short-run marginal price elasticity between -0.18 and -0.36 whereas the estimated long-run marginal price elasticity varied from -0.27 to -0.71.

We show in Section 2 of this paper that it is possible to derive a dynamic relation in consumption terms from a structural optimisation program solved by the municipality. We assume in this program that the community has a twofold objective: the maximisation of local consumers' utility and the reduction in the debt of its water service. The control variable is the price of water charged to the domestic users. When solving for this intertemporal program, we obtain a nonlinear dynamic equation, in which the variation in consumption between two successive periods depends on the variation in prices and on the income of the representative consumer. This is the consumption path which corresponds to the prices chosen by the local authority. In other words, we obtain the water supply function of the community. The local authority takes explicitly into account that the domestic consumer responds to prices at each time period, through the specification of the indirect utility function. Thus, we can infer that demand is going to match supply and that the dynamic relation obtained after solving the program corresponds also to the demand function. This last result will allow us to estimate price elasticities of demand.

We show next that this model can be linearised. The necessary assumptions for linearisation describe the annual adjustment of price in the municipalities and assume some heterogeneity between the communities. As a result, current consumption is written as a linear function of lagged consumption, current price and income. While considering a cross-section of communities, these conditions also lead to a particular specification of the error term, originating in particular from the panel data structure. It is commonly the case that a term of individual heterogeneity is added in panel data models, this term being time-invariant (Anderson and Hsiao, 1982; Arellano and Bond, 1991; Ahn and Schmidt, 1995, 1997) or affected by a multiplicative time-varying parameter (Holtz-Eakin et al., 1988). The linear relation we obtain combines these two terms of unobserved heterogeneity in a mixed error structure.

In Section 3 of this paper, we focus on econometric methods referring largely to the literature on dynamic panel data models (Arellano and Bond, 1991; Ahn and Schmidt, 1995, 1997; Blundell and Bond, 1998). We show first that the combined error structure of the dynamic linear model requires original estimation methods. Indeed, if the mixed structure of our model is the true data generating process, we show that the usual Generalized Method of Moments (GMM) estimation procedures based on first-difference (FD) and quasi-difference (QD) transformations and using as instruments the lagged dependent variable do not generally produce consistent estimates. So, we propose a new estimator combining the FD and QD transformations and we describe the corresponding orthogonality conditions. We show that this estimator is consistent and efficient in the case of the mixed error term structure. Moreover, if the correct specification includes only a purely time-invariant effect or an individual effect affected by a temporal parameter, the new estimator will still be consistent but no longer efficient.

In Section 4, we estimate the two dynamic models: the nonlinear model directly derived from the optimisation program and the linear model with the mixed error structure. The estimation is conducted on a sample of French communities observed during the 1988–1993 period. Hansen tests reveal that the correct specification of the linear model is the mixed error structure which combines the two terms of unobserved heterogeneity. We next show, using a test proposed by Rivers and Vuong (1991), that the two models are asymptotically equivalent.

We find a long-run (on a six-year period) price elasticity (−0.40) higher, in absolute value, than the short-run price elasticity (−0.26). This result is quite important for the water authorities. It implies that the planning board should be careful when assessing the impact of variation in prices on consumption and well-being.

Section 5 concludes.

2. Where Does the Dynamic Demand Function Come From?

2.1. Some Elements on the Management of Water Services in France

In France, water supply and wastewater treatment services are the responsibility of communities. The municipalities have the choice between managing the water utilities by themselves (through a local, public water authority) or delegating their operation to a private company.[1] Whatever the form of the management, the water network (pipelines, service connections, reservoirs, treatment facilities, etc.) is owned by the municipality. We focus in this paper on the case of a community which has decided to delegate the management of its water utility to a private operator.

Once the operator has been chosen (through a public bid procedure), the relationship between the municipality and the company is formalised by means of a contract. At the beginning of the contract period, the community and the private company have to decide by mutual consent on the price that will be charged to the consumers. It is commonly the case that a degree of price discrimination is used among broad categories of users: municipal, domestic, industrial and agricultural. We focus here on the determination of the price of water for domestic users. When the private company is operating the water utility, the community remains in charge of the financing of the water system (this kind of leasing is the most common mode of delegation in France). A part of the price charged to customers is intended to remunerate the private operator and the other part is meant to cover the investment charges borne by the municipality (we will call this second part the 'municipal charge').

Let us consider a community which has decided to delegate the management of the service of water supply and/or the service of wastewater treatment to a private operator. We suppose that the company has already been chosen and that the local representatives are about to negotiate the price that will be charged to their domestic customers. In the next paragraph, we describe the calculation made by the local representatives in order to determine the price they will propose to the private operator.[2]

2.2. Intertemporal Model of Price Determination

We assume that the local representatives solve a partial optimisation problem focusing only on water supply. The municipality has a twofold objective: maximising the utility of the domestic users and recovering the debt of the service of water (since the French water act in 1992, the local governments have to manage separately the budget of water and the overall budget of the municipality). The indirect utility function is assumed to depend only on the price of water and the income. The local authority chooses the water price series in order to maximise the discounted difference between the indirect utility of the representative consumer and the average debt per user.

The program starts at time $t = 0$ and is written as:

$$\max_{\{P_t\}} \sum_{t=0}^{\infty} (1+\rho)^{-t} [v(P_t, M_t) - D_t]$$

$$\text{under the constraints} \begin{cases} D_{t+1} = D_t(1+r) - kP_tC_t, \\ D_0 = \bar{D}, \\ P_0 = \bar{P}. \end{cases}$$

r is the interest rate and ρ is the discount rate. The initial price P_0 and debt D_0 are assumed known. D_t is the average water utility debt per user. P_t and C_t are respectively the price of water and the annual domestic water consumption in period t. We assume also that the local authority takes explicitly into account the impact of water prices on domestic users' consumption, forecasting C_t using the usual functional form:

$$C_t = BP_t^{\beta_1} M_t^{\beta_2} \quad \text{where} \quad B = e^{\beta_0} \sum_{j=3}^{m} Z_j^{\beta_j}.$$

Z_j ($j = 3, \ldots m$) are the variables describing the user and its environment, in addition to the price and income variables. β_j ($j = 0, \ldots m$) are unknown parameters. The β_1 and β_2 parameters are the constant price and income elasticities. It is easy to show, using duality conditions, that this Marshallian demand function corresponds to the following indirect utility function:

$$v(P_t, M_t) = \frac{M_t^{1-\beta_2}}{1-\beta_2} - B\frac{P_t^{1+\beta_1}}{1+\beta_1}.$$

The first constraint in the program means that a share k of the total water utility bill ($P_t C_t$) is intended to reimburse the debt. This k parameter will be exogenously fixed because it is almost constant across communities.

We can write the Bellman equation as:

$$V_t(D_t) = \max_{\{P_t\}}[v(P_t, M_t) - D_t + (1+\rho)^{-1} V_{t+1}(D_{t+1})].$$

We have the following first-order condition:

$$\frac{\partial V_t}{\partial P_t} = 0 \Leftrightarrow \frac{\partial v(P_t, M_t)}{\partial P_t} + (1+\rho)^{-1}\frac{\partial V_{t+1}}{\partial D_{t+1}}[-k(1+\beta_1)C_t] = 0, \tag{1}$$

and from the envelope theorem:

$$\frac{\partial V_t}{\partial D_t} = -1 + (1+\rho)^{-1}\frac{\partial V_{t+1}}{\partial D_{t+1}}(1+r). \tag{2}$$

Condition (1) can then be written:

$$-BP_t^{\beta_1} + (1+\rho)^{-1}\frac{\partial V_{t+1}}{\partial D_{t+1}}[-k(1+\beta_1)C_t] = 0. \tag{3}$$

From (3), we have:

$$\frac{\partial V_{t+1}}{\partial D_{t+1}} = \frac{-BP_t^{\beta_1}(1+\rho)}{k(1+\beta_1)C_t} \tag{4}$$

which, using (2), leads to

$$\frac{\partial V_t}{\partial D_t} = -1 - \frac{BP_t^{\beta_1}(1+r)}{k(1+\beta_1)C_t}.$$

Lagging Equation (3) and introducing Equation (4), we get:

$$-BP_{t-1}^{\beta_1} + (1+\rho)^{-1}\left[-1 - \frac{BP_t^{\beta_1}(1+r)}{k(1+\beta_1)C_t}\right][-k(1+\beta_1)C_{t-1}] = 0.$$

Dividing by C_{t-1} and arranging terms, we have:

$$-B\frac{P_{t-1}^{\beta_1}}{C_{t-1}} + (1+\rho)^{-1}(1+\beta_1)k + (1+\rho)^{-1}(1+r)\frac{BP_t^{\beta_1}}{C_t} = 0.$$

Multiplying the two terms of the last equation by $C_t/P_t^{\beta_1}$, we get:

$$-B\frac{P_{t-1}^{\beta_1}}{C_{t-1}}\frac{C_t}{P_t^{\beta_1}} + (1+\rho)^{-1}(1+\beta_1)k\frac{C_t}{P_t^{\beta_1}} + (1+\rho)^{-1}(1+r)B = 0$$

$$\Leftrightarrow\; B\frac{C_t}{C_{t-1}}\frac{P_{t-1}^{\beta_1}}{P_t^{\beta_1}} = \frac{k(1+\beta_1)}{(1+\rho)}\frac{C_t}{P_t^{\beta_1}} + \frac{(1+r)}{(1+\rho)}B.$$

Using the relation $C_t = BP_t^{\beta_1}M_t^{\beta_2}$ and multiplying by $(1+\rho)/(1+r)$ leads to:

$$\frac{C_t}{C_{t-1}}\left(\frac{1+\rho}{1+r}\right) = \left(\frac{P_t}{P_{t-1}}\right)^{\beta_1}\left(\frac{k(1+\beta_1)M_t^{\beta_2}}{1+r} + 1\right)$$

$$\Leftrightarrow \log\left(\frac{C_t}{C_{t-1}}\right) = -\log\left(\frac{1+\rho}{1+r}\right) + \beta_1\log\left(\frac{P_t}{P_{t-1}}\right) + \log\left(\frac{k(1+\beta_1)M_t^{\beta_2}}{1+r} + 1\right). \tag{5}$$

This equation relates the ratio between consumption at time t and consumption at time $t-1$ to the ratio of prices between the same time periods, income at time t, the discount rate ρ and the interest rate r. This relation provides the dynamics of consumption as a result of the municipality's optimisation program. This is the consumption path which corresponds to the price series chosen by the local authority. For this reason, we can call it the 'municipality supply function'. As the policy maker takes explicitly into account the impact of prices on the consumer's behaviour (through the indirect utility function in its program), we can say that this function is also exactly the demand function of the representative user. So, it will allow us to compute price elasticities of demand.

This relation is nonlinear in the parameters. We show in the next section that this equation can be easily linearised, making assumptions about the annual adjustment of prices and assuming heterogeneity between the communities.

2.3. Linearisation of the Dynamic Model

We recall that, for local community i, the optimal solution of the program is the following dynamic relation:

$$\log\left(\frac{C_{it}}{C_{i,t-1}}\right) = -\log\left(\frac{1+\rho}{1+r}\right) + \beta_1 \log\left(\frac{P_{it}}{P_{i,t-1}}\right) + \log\left(\frac{k(1+\beta_1)M_{it}^{\beta_2}}{1+r} + 1\right). \quad (6)$$

Model (6) can be rewritten as:

$$\log\left(\frac{C_{it}}{C_{i,t-1}}\right) = -\log(1+\rho) + \log(1+r) + \beta_1 \log(P_{it}) - \beta_1 \log(P_{i,t-1}) + \log\left(\frac{k(1+\beta_1)M_{it}^{\beta_2}}{1+r} + 1\right)$$

$$\Leftrightarrow \log\left(\frac{C_{it}}{C_{i,t-1}}\right) = \beta_1 \log(P_{it}) + \log(k(1+\beta_1)M_{it}^{\beta_2} + (1+r)) - \log(1+\rho) - \beta_1 \log(P_{i,t-1}).$$

It is common practice, when negotiating for the price of water at the beginning of the contract, that the private company and the community agree on the initial price and on an annual updating rule. More precisely, the price of water is updated each year through a multiplicative coefficient a_t which depends on public reports of input price indexes (labor cost, construction cost, energy, etc.). The relation between prices can then be written as:

$$P_{it} = P_{i,t-1}(1+a_t) = P_{i,t-2}(1+a_{t-1})(1+a_t) = \ldots = P_{i0}\prod_{s=1}^{t}(1+a_s),$$

where P_{i0} is the price negotiated at the beginning of the contract. Choosing P_{it} is thus equivalent to choosing P_{i0} first and updating terms a_t accordingly. We simplify this last equation in the following manner, noting that, for a_s small, $1 + a_s \simeq \exp(a_s)$:

$$P_{i0}\prod_{s=1}^{t}(1+a_s) \simeq P_{i0}\exp\left(\sum_{s=1}^{t} a_s\right) = v_i\theta_t$$

with $\theta_t = \exp\left(\sum_{s=1}^{t} a_s\right)$ for $t \geq 1$ and $v_i = P_{i0}$.[3]

We will assume also that there exists some heterogeneity between communities: local governments are supposed to behave in the same way (i.e. to solve the same optimisation program) but to differ in terms of intertemporal preferences. So, we define a new random variable, ρ_i, with mean $\bar{\rho}$ ($\bar{\rho} \neq 0$), which is the discount rate for community i.

It remains a nonlinear term in the equation:

$$\log[k(1+\beta_1)M_{it}^{\beta_2} + (1+r)],$$

where k and r are constant and β_1 and β_2 are respectively price and income elasticities. For reasonable values of the parameters ($r = 0.05$, $\beta_1 = -0.2$ and $\beta_2 = 0.1$) and for the values of the municipal share k ($k = 0.16$ in average) and the income variable in our sample, we check that:

$$\beta_2 \log(M_{it}) \simeq \log(k(1+\beta_1)M_{it}^{\beta_2} + (1+r)).$$

We may now introduce the usual error term ε_{it}, assumed identically and independently distributed across communities and time periods. Under these assumptions, model (6) becomes:

$$\log\left(\frac{C_{it}}{C_{i,t-1}}\right) = \beta_1 \log(P_{it}) + \log[k(1+\beta_1)M_{it}^{\beta_2} + (1+r)] + \theta_t v_i + \alpha_i + \varepsilon_{it}$$

$$\text{with} \quad \alpha_i = \log(1+\bar{\rho}) - \log(1+\rho_i).$$

Under this whole set of assumptions, and relaxing the unit root hypothesis (a parameter δ will measure the effect of the lagged consumption variable), the intertemporal model can be rewritten as:

$$\log(C_{it}) = \delta \log(C_{i,t-1}) + \beta_1 \log(P_{it}) + \beta_2 \log(M_{it}) + \alpha_i + \theta_t v_i + \varepsilon_{it}. \quad (7)$$

In summary, we have shown that, under some assumptions, the dynamic equation derived from the structural optimisation program can be written as a dynamic panel data model of consumption, linear in the parameters, but with a particular error term structure combining two terms of unobserved individual heterogeneity.

3. Econometric Methods

3.1. Usual Estimation Procedures

In the usual dynamic panel data model, endogeneity bias typically arises because of the joint presence of the individual effect and the lagged dependent variable in the right-hand side of the equation to be estimated (Hsiao, 1986). Furthermore, when incorporating additional regressors, potential correlation between some of those and the individual effect is another source of inconsistency of least squares estimates. These two sources of endogeneity bias are traditionally addressed in the literature by building a set of orthogonality conditions and estimating the model with GMM (Generalized Method of Moments) (Hansen, 1982).

The individual effect can be time-invariant (Arellano and Bond, 1991; Ahn and Schmidt, 1995, 1997; Blundell and Bond, 1998) or allowed to vary across time periods (Holtz-Eakin et al., 1988). The individual effect is commonly eliminated by first-differencing the model in the former case and by a quasi-differentiation technique in the latter. Then, instrument matrices can be constructed from dependent variables lagged two periods or more (Arellano and Bond, 1991; Ahn and Schmidt, 1995, 1997).

Dynamic panel data models are commonly written as:

$$Y_{it} = \delta Y_{i,t-1} + X'_{it}\beta + u_{it}, \quad i = 1, \dots, N; t = 1, \dots, T.$$

Initial values Y_{i0} are assumed known and possibly stochastic. Individual heterogeneity is accounted for by decomposing u_{it} into two terms: either $\mu_i + \eta_{it}$ when the individual effect is assumed time-invariant, or $\theta_t v_i + \varepsilon_{it}$ when the individual effect is allowed to vary across time periods.

3.1.1. *Purely Time-Invariant Individual Effect*

The model is the following:

$$Y_{it} = \delta Y_{i,t-1} + X'_{it}\beta + u_{it}$$

$$\text{with} \quad u_{it} = \mu_i + \eta_{it}, \quad i = 1, \dots, N; t = 1, \dots, T.$$

η_{it} is the usual econometric error term, assumed independently and identically distributed. The random variables μ_i and η_{it} are assumed to have both zero mean and not to be correlated ($E(\mu_i \eta_{it}) = 0$). The X_{it} variables are supposed strictly exogenous in the sense that:

$$E(X_{it}\eta_{is}) = 0 \quad \forall t, s,$$

but we allow these variables to be correlated with the individual error term μ_i.

If $T \to \infty$, estimating the model where the variables have been deviated from their means, by Ordinary Least Squares (OLS), produces consistent estimates. It is commonly the case that the period covered in panel data samples is rather short and other estimation techniques have to be considered. The solution usually adopted is to first-difference the model in order to eliminate the individual effect:

$$\begin{aligned} Y_{it} - Y_{i,t-1} &= \delta(Y_{i,t-1} - Y_{i,t-2}) + (X'_{it} - X'_{i,t-1})\,\beta + \eta_{it} - \eta_{i,t-1} \\ \Leftrightarrow \quad \Delta Y_{it} &= \delta \Delta Y_{i,t-1} + \Delta X'_{it}\,\beta + \Delta\eta_{it}, \end{aligned}$$

where Δ is the first-difference operator. It is now possible to use as instruments realisations of the lagged dependent variable, following Anderson and Hsiao (1982) or Arellano and Bond (1991). Arellano and Bond (1991) show that, under the assumption of no serial correlation between the η_{it} error terms, the dependent variable lagged two periods or more is a valid instrument for the equation in first-difference. So, for each individual i, we have the following $m = T(T-1)/2$ moment conditions:

$$E(Y_{is}\Delta u_{it}) = 0 \quad \text{for} \quad t \geq 2 \quad \text{and} \quad s = 0, \ldots, t-2. \tag{8}$$

The strictly exogenous variables can also be added in the matrix of instruments and GMM estimation can be performed.

3.1.2. *Individual Effect and Time-Varying Exogenous Shocks*

The model is the following:

$$Y_{it} = \delta Y_{i,t-1} + X'_{it}\beta + u_{it}$$

$$\text{with} \quad u_{it} = \theta_t v_i + \varepsilon_{it}, \quad i = 1, \ldots, N; t = 1, \ldots, T.$$

ε_{it} is assumed independently and identically distributed. The random variables v_i and ε_{it} are assumed to have both zero mean and not to be correlated. In order to eliminate the individual effect, we use the quasi-difference transformation along the lines of Holtz-Eakin et al. (1988). The procedure is to subtract from the model at time t the lagged equation multiplied by $r_t = \theta_t/\theta_{t-1}$. The transformed model of demand is thus:

$$Y_{it} - r_t Y_{i,t-1} = \delta(Y_{i,t-1} - r_t Y_{i,t-2}) + (X'_{it} - r_t X'_{i,t-1})\beta + \varepsilon_{it} - r_t \varepsilon_{i,t-1}.$$

The parameters to be estimated are $(\delta, \beta, r_t, t = 2, \ldots, T)$.

The orthogonality conditions become $E(W'_i\widetilde{u}_i) = 0$ where W_i is the matrix of instruments for individual i and $\widetilde{u}_i = (\widetilde{u}_{i2}, \ldots, \widetilde{u}_{iT})'$ with $\widetilde{u}_{it} = \varepsilon_{it} - r_t\varepsilon_{i,t-1}$. The number of moment conditions including the lagged dependent variable, equivalent to the ones proposed by Arellano and Bond (1991), is the same as in the previous case: $T(T-1)/2$. It is still the case that the estimators will be consistent if and only if the residuals are not first-order serially correlated.

3.2. Inconsistency of Usual Estimation Procedures

Nauges and Thomas (1999) consider a model with mixed covariance structure, including the two terms of individual heterogeneity (α_i and v_i), which can have different distributions and can be correlated:

$$Y_{it} = \delta Y_{i,t-1} + X'_{it}\beta + u_{it}$$

$$\text{with} \quad u_{it} = \alpha_i + \theta_t v_i + \varepsilon_{it}, \quad i = 1, \ldots, N; t = 1, \ldots, T.$$

Some additional assumptions are made:

ε_{it} is i.i.d. between individuals and between time periods,

$$E(\alpha_i^2) = \sigma_\alpha^2, \quad E(v_i^2) = \sigma_v^2, \quad E(\varepsilon_{it}^2) = \sigma_\varepsilon^2 \quad \forall i, \forall t,$$

$$E(\varepsilon_{it}\alpha_i) = E(\varepsilon_{it}v_i) = 0, \quad E(Y_{i0}\varepsilon_{it}) = 0 \quad \forall t, \quad E(\alpha_i v_i) = \sigma_{\alpha v}. \tag{9}$$

$Y_{i,0}$ ($i = 1, \ldots, N$) are initial conditions, assumed known and possibly stochastic. This general error term structure includes as special cases the two models described previously. It is easy to show that under the two following conditions:

$$\theta_t = \theta_s \quad \forall t, s \tag{10}$$

$$\sigma_{\alpha v} = \sigma_v^2 = 0 \quad \text{with} \quad \sigma_v^2 = E(v_i^2), \tag{11}$$

the model with mixed covariance structure reduces to the model with purely stationary individual effect where the error term is specified as $u_{it} = \alpha_i + \varepsilon_{it}$.

Let us now consider the case where the individual effects α_i and v_i are perfectly correlated:

$$\rho_{\alpha v} = \{-1, 1\}. \tag{12}$$

In this case, the mixed error structure reduces to Holtz et al.'s model where the individual effect is purely nonstationary.

Nauges and Thomas (1999) show that transforming the model with the mixed error structure by the usual techniques of first-differencing and quasi-differencing does not eliminate both individual error terms. When operating the first-difference transformation on the mixed error term, we obtain the following transformed residual:

$$u_{it} - u_{i,t-1} = (\theta_t - \theta_{t-1})v_i + \varepsilon_{it} - \varepsilon_{i,t-1}$$

which still depends on v_i. More precisely, we have

$$E[(u_{it} - u_{i,t-1})Y_{is}] = (\theta_t - \theta_{t-1})\sigma_{\alpha v} + \theta_s(\theta_t - \theta_{t-1})\sigma_v^2, \quad s \leq t-2.$$

When quasi-differencing the model, the transformed residual is:

$$u_{it} - r_t u_{i,t-1} = (1 - r_t)\alpha_i + \varepsilon_{i,t} - r_t\varepsilon_{i,t-1} \quad \text{with} \quad r_t = \theta_t/\theta_{t-1}$$

which still depends on α_i. We have

$$E[(u_{it} - r_t u_{i,t-1})Y_{is}] = (1 - r_t)\sigma_\alpha^2 + (1 - r_t)\theta_s\sigma_{\alpha v}, \quad s \leq t-2.$$

So, if the model is transformed by the usual techniques of first-differencing and quasi-differencing, the orthogonality conditions including the lagged dependent variable are no longer valid (except if one of the conditions (10), (11) or (12) is verified) because the correlation between the individual error term and the lagged dependent variable has not been eliminated. Therefore, the GMM estimators based on these orthogonality conditions will not be consistent.

3.3. A Consistent Transformation Method

Nauges and Thomas (1999) propose a consistent double (two-step) transformation of the dynamic model with the mixed error structure in order to eliminate both individual effects. They first apply a first-difference transformation. The α_i individual effect is eliminated and the transformed residual is written $\Delta\theta_t v_i + \Delta\varepsilon_{it}$. If we subtract the differenced equation of the previous period, multiplied by the coefficient $\Delta\theta_t/\Delta\theta_{t-1}$, we get:

$$\Delta Y_{it} - \tilde{r}_t \Delta Y_{i,t-1} = \delta(\Delta Y_{i,t-1} - \tilde{r}_t \Delta Y_{i,t-2}) + (\Delta X'_{it} - \tilde{r}_t \Delta X'_{i,t-1})\beta$$
$$+ \Delta\varepsilon_{it} - \tilde{r}_t \Delta\varepsilon_{i,t-1}, \quad i = 1, 2, \ldots, N; t = 2, 3, \ldots, T,$$

and

$$\tilde{r}_t = \Delta\theta_t/\Delta\theta_{t-1} = (\theta_t - \theta_{t-1})/(\theta_{t-1} - \theta_{t-2}).$$

Our estimator can be seen as a GMM on double-differenced data. Given the conditions above on α_i, v_i, and ε_{it}, it is easy to show that the following restrictions must be valid :

$$E\left[Y_{is}(\Delta u_{it} - \tilde{r}_t \Delta u_{i,t-1})\right] = 0 \quad t = 2, \ldots, T; s = 0, \ldots, t-2. \tag{13}$$

We have $(T-1)(T-2)/2$ such conditions, which are equivalent to those in Arellano and Bond (1991). These moment conditions are nonlinear in the parameters, as $\tilde{r}_t$'s are estimated jointly with the parameter of interest (δ). We verify that the GMM estimator based on conditions (13) is efficient by comparing the number of conditions obtained above to the number of conditions identified from covariance restrictions (see Nauges and Thomas, 1999).

3.4. Specification Tests

We described in the preceding section three GMM estimators which are based on different sets of orthogonality conditions. Specification tests can be considered, following Hausman's (1978) approach. If the error term is made of a unique term of individual heterogeneity (purely time-invariant or affected by a temporal shock), the estimator based on the double difference is still consistent but will not be efficient. This is because it does not exploit all moment conditions available in this case. Indeed, compared to the First-Difference (FD) and the Quasi-Difference (QD) estimator, we have less active moment conditions ($(T-1)(T-2)/2$ instead of $T(T-1)/2$), when the matrix of instruments consists of lagged dependent variables only. That is, we have T additional conditions when using First-Difference or Quasi-Difference. Table 1 summarizes the properties of these three estimators in terms of consistency and efficiency, depending on the structure of the individual heterogeneity.

Table 1. Properties of the three GMM estimators.

Type of individual heterogeneity	FD estimator[1]	QD estimator[2]	DD[3] estimator
purely time-invariant α_i	consistent and efficient	not consistent	consistent and not efficient
affected by a temporal shock $\theta_t v_i$	not consistent	consistent and efficient	consistent and not efficient
mixed structure $\alpha_i + \theta_t v_i$	not consistent	not consistent	consistent and efficient

[1]GMM estimation method based on a First-Differentiation of the model.
[2]GMM estimation method based on a Quasi-Differentiation of the model.
[3]GMM estimation method based on a Double-Differentiation of the model.

Table 2. Descriptive statistics on the sample.

Variable	Mean	St. Dev.	Minimum	Maximum
consumption[1] (m^3)	152.11	40.09	70.00	334.00
price of water (FF/m^3)	8.64	3.29	3.34	20.66
income (FF)	105316	18466	53598	324661

Number of observations: 696 ($N = 116$ and $T = 6$).
[1]Annual water consumption of the representative domestic user.

4. Description of the Data Base and Estimation Results

4.1. Data Description

We use a panel data sample of 116 communities from Eastern France, on the period 1988–1993. These 116 municipalities are supplied by the Compagnie Générale des Eaux (now Vivendi), a major private operator. This company provided us with data on aggregate residential water consumption and on the price of water in each community. We also know the number of domestic accounts so we can easily obtain the annual water consumption (C_{it}, in cubic meter) of the representative domestic user in each community. In the area considered, residential users are facing a two-part tariff. They have to pay a fixed charge corresponding to the connection fee and a price per unit of water consumed. The fixed charge corresponds to fixed costs supported by the operator, while the price per unit corresponds broadly to operation and maintenance costs proportional to water actually supplied. For each of the 116 communities and for each year of the 1988–1993 period, information on income was obtained from the French National Institute of Statistics and Economic Studies (INSEE). Summary statistics for these variables are presented in Table 2.

Table 3. GMM estimation of the nonlinear dynamic model.

Parameter	Coefficient	Standard Error
β_0	0.0280	0.1099
β_1	–0.2646	0.0897**
β_2	–0.3366	0.9725

$N = 116$ and $T = 6$.
**Significant at the 1% level.

4.2. Estimation of the Nonlinear Model

The model directly derived from the optimisation program was the following:

$$\log\left(\frac{C_{it}}{C_{i,t-1}}\right) = -\log\left(\frac{1+\rho}{1+r}\right) + \beta_1 \log\left(\frac{P_{it}}{P_{i,t-1}}\right) + \log\left(\frac{k(1+\beta_1)M_{it}^{\beta_2}}{1+r} + 1\right) + u_{it}. \tag{14}$$

$$i = 1, \ldots, N; \quad t = 1, \ldots, T.$$

We add the usual econometric error term u_{it} assumed of mean zero. C_{it} is the consumption of the representative domestic user in community i at time t. r and k are assumed known by the community. We choose $k = 0.16$ which is the average value in the sample and $r = 0.05$. The parameters to be estimated are thus β_1, β_2 and ρ. We will denote β_0 the $[\log(1+r) - \log(1+\rho)]$ term. This model is estimated using the GMM two-step estimator, to control for possible heteroskedasticity in the residual term u_{it}. The instruments chosen are a constant, the current and the lagged price and the lagged amount of the municipal charge.

We test for the validity of the instruments using a Hansen test. Under the null hypothesis of valid instruments, the GMM criterion is distributed as a chi-square statistic with l degrees of freedom, l being the number of overidentifying constraints (Hansen, 1982). In this model, $l = 1$. The GMM criterion is equal to 0.0008, which is less than the critical value of a chi-square distribution with one degree of freedom. Estimation results are presented in Table 3.

The Student test associated with the β_0 parameter in the model amounts to test if $((1+\rho)/(1+r))$ is significantly different from 1 or, equivalently, if ρ is significantly different from r. The Student test statistic does not lead to the rejection of the null that r and ρ are equal. r being fixed at 0.05, we can find the estimated value of ρ:

$$\hat{\rho} = e^{\hat{\beta}_0}(1+r) - 1 \Rightarrow \hat{\rho} = e^{0.0280}(1.05) - 1 = 0.08.$$

The discount factor is estimated at 0.08. The price coefficient is significant at the 1% level and estimated at –0.26. This figure corresponds to the short-run price elasticity of demand, and is close to the ones found in other European empirical studies on residential water demand (Point, 1993; CREDOC, 1997; Höglund, 1997). The coefficient of the income variable, which corresponds to the income elasticity, is not significant.

4.3. Estimation of the Linear Dynamic Model

The dynamic model of residential water consumption to be estimated is the following:

$$C_{it} = \delta C_{i,t-1} + \beta_1 P_{it} + \beta_2 M_{it} + \alpha_i + \theta_t v_i + \varepsilon_{it}, \quad i = 1, \ldots, N;\ t = 1, \ldots, T,$$

where all variables are in logarithmic form. This model is estimated by GMM using Blundell and Bond's (1998) orthogonality conditions. These authors propose, in the case of a purely time-invariant term of individual heterogeneity, the use of an extended linear GMM estimator that uses lagged differences of the dependent variable as instruments for equations in levels, in addition to lagged levels of the dependent variable as instruments for equations in first differences (see also Arellano and Bover, 1995). The Blundell and Bond's $(T-4)$ extra conditions for the model with the two terms of individual heterogeneity are:

$$E[u_{it}(\Delta C_{i,t-1} - \widetilde{r}_{t-1}\Delta C_{i,t-2})] = 0, \quad t = 3, \ldots, T. \tag{15}$$

Moreover, if the ε_{it} disturbances are homoskedastic through time, i.e. if

$$E(\varepsilon_{it}^2) = \sigma_i^2 \quad \text{for} \quad t = 1, \ldots, T,$$

we have $(T-3)$ extra orthogonality restrictions of the form (Ahn and Schmidt, 1995, 1997):

$$E[C_{i,t-2}(\Delta u_{i,t-1} - \widetilde{r}_{t-1}\Delta u_{i,t-2}) - C_{i,t-1}(\Delta u_{it} - \widetilde{r}_t\Delta u_{i,t-1})] = 0, \tag{16}$$

where $t = 4, \ldots, T$.

We compare the results obtained on our sample with the First-Difference, the Quasi-Difference and our Double-Difference estimator. We test for the validity of the Blundell and Bond's orthogonality conditions using a Hansen test. The results are presented in Table 4.

The Hansen test rejects the null of valid orthogonality conditions at the 5% level in two cases out of three. The only consistent estimator is the one based on the Double-Difference transformation (p-value: 0.0504). So, it seems that the correct specification of the error term is the one with the two terms of individual heterogeneity. The price of water has a significant negative effect on water consumption whereas income has a significant positive impact. The autoregressive parameter is estimated at 0.19. The long-run price [resp. income] elasticities can be computed using the following formula:[4]

$$\varepsilon = \frac{\beta}{1-\delta},$$

Table 4. Estimation results (GMM based on Blundell and Bond's orthogonality conditions).

Variable	FD estimator		QD estimator		DD estimator	
	Coefficient	St. Err.[(1)]	Coefficient	St. Err.	Coefficient	St. Err.
lagged consumption	0.3683**	0.0820	0.6665**	0.0890	0.1939**	0.0866
price of water	–0.2530**	0.0236	–0.0893**	0.0267	–0.3186**	0.0264
income	0.3200**	0.0390	0.0152**	0.0052	0.4080**	0.0410
r_{1990}	...	...	0.7669**	0.1208	...	...
r_{1991}	...	...	0.8962**	0.1307	–0.8078*	0.4051
r_{1992}	...	...	2.2138**	0.2151	–0.5985	0.3703
r_{1993}	...	...	0.3737**	0.0322	–0.3517	0.2405
Hansen test statistic (t)	44.1500		62.4726		23.6591	
Degrees of freedom	22		18		11	
Prob > t	0.0034		0.0000		0.0504	

FD: First-Difference, QD: Quasi-Difference, DD: Double-Difference.
(1)Standard Error.
...: not identifiable.
*significant at the 5% level; **significant at the 1% level.

where β is the coefficient of the price [resp. income] and δ is the coefficient of the lagged dependent variable. Using the estimation results of Table 4, we obtain the following results:

$$\varepsilon_{\text{price}} = -0.40 \quad \text{and} \quad \varepsilon_{\text{inc.}} = 0.51.$$

The short-run price elasticity of demand has been estimated previously at -0.26 (see Table 3). On this six-year period, the long-run price elasticity is about 1.5 times higher than the short-run price elasticity. The explanation could be that the price of water becomes a signal for the consumer only after a persistent increase in tariffs. It seems that it takes more than one year for the consumer's behaviour to adapt to the variation in prices. One reason could be that prices in France have been very low since a long time. They increased sharply at the beginning of the nineties. People could have got into bad habits at that time, wasting water for example. And it probably takes a few years to get out of them. Moreover, there are more and more information campaigns promoting water conservation in France. Consumers could have been sensitive but the reaction is also commonly slow.

Although the long-run price elasticity remains, in absolute value, less than one, this figure is greater than the ones estimated in other European studies, in static framework. Hansen (1996), Höglund (1997) and Point (1993) found a short-run price elasticity respectively equal to -0.10 for Denmark, -0.20 for Sweden, and -0.17 for the South of France. The higher long-run price elasticity is a quite important result because it states that households are more likely to adjust their consumption after a permanent increase in tariffs than after a one year shift in the price of water. This should be taken in consideration by policy makers when forecasting the impact of price increase on consumption and well-being. Long-run measures of elasticity should be preferred to short-run ones which underestimate the reaction of households.

Moreover, on a longer period, households could replace some of their durable equipment. And we know that water is consumed mainly through sanitary and housing fittings (washing machine, dishwasher). The water consumption of these equipment is becoming more and more considered by households at the time of purchasing. Replacement by some equipment less water consuming could lead to a more important decrease in consumption. In other words, long-run price elasticity (on a time period longer than six years) could be even greater.

We also note that income has a significant effect on water consumption, all other things being equal. This result indicates that local governments may find useful to impose progressive tariffs for water. Increasing tariffs can be a mean of redistributing income from consumers with a greater water consumption towards consumers with a more modest consumption.

We note that the estimation of β_1 is quite close in the two models: -0.26 in the nonlinear model (see Table 3) and -0.32 in the linear one (see Table 4). In the following section, we test for the asymptotic equivalence between the two models using a test proposed by Rivers and Vuong (1991).

4.4. Test of Asymptotic Equivalence

The basic idea of this procedure is to test for the significance of the difference of the sample lack-of-fit criteria for competing models. Here, along the lines of Carpentier and Weaver (1997), we choose as a test statistic a function of the empirical moment conditions. So, we measure the lack-of-fit criterion by the sum of the squared estimated values of their deviations from zero, i.e.:

$$Q(\hat{\beta}) \equiv M(\hat{\beta})' M(\hat{\beta}),$$

where

$$M(\hat{\beta}) \equiv \frac{1}{N} \sum_{i=1}^{N} w_i(\hat{\beta})' \bar{u}_i(\hat{\beta}).$$

w_i and $\bar{u}_i$ are respectively the instruments and the transformed residual for individual i.

We will note these sums for the nonlinear model and the linear one by $Q_1(\hat{\beta}^1)$ and $Q_2(\hat{\beta}^2)$ respectively.

The Rivers–Vuong test statistic is defined as a normalised difference of the sample lack-of-fit criteria:

$$\hat{T} \equiv \frac{\sqrt{N}}{\hat{\sigma}} [Q_1(\hat{\beta}^1) - Q_2(\hat{\beta}^2)],$$

where $\hat{\sigma}$ is a consistent estimator of the asymptotic standard deviation of the difference of lack-of-fit.

Under the regularity conditions defined by Rivers and Vuong, we examine three competing nonnested hypothesis by comparing the asymptotic lack-of-fit of the competing models. The null hypothesis H_0 is that the two models are asymptotically equivalent when

$$H_0 : \lim_{N\to\infty} [Q_1(\hat{\beta}^1) - Q_2(\hat{\beta}^2)] = 0.$$

The first alternative hypothesis is that the nonlinear model (here called model 1) is asymptotically better than the linear model (called model 2):

$$H_1 : \lim \sup_{N\to\infty} [Q_1(\hat{\beta}^1) - Q_2(\hat{\beta}^2)] < 0.$$

Similarly, the second alternative hypothesis is that the linear model (model 2) is asymptotically better than the nonlinear one (model 1):

$$H_2 : \lim \inf_{N\to\infty} [Q_1(\hat{\beta}^1) - Q_2(\hat{\beta}^2)] > 0.$$

The test procedure involves comparing values of the estimated statistic with critical values of a standard normal distribution. Given a significance level of 5%, H_1 cannot be rejected against H_0 if

$\hat{T} < -1.96$, H_2 cannot be rejected against H_0 if $\hat{T} > 1.96$, and H_0 is not rejected otherwise. In our case, the estimated Rivers–Vuong statistic is 0.002 so we cannot reject the null of asymptotic equivalence between the two models. This result means that the conditions imposed on the nonlinear model do not modify significantly the estimation results.

5. Conclusion

We show in this paper that a dynamic model of residential water consumption can be derived from a structural optimisation program solved by the communities. This nonlinear model proves asymptotically equivalent to more common dynamic panel data models that are linear in the parameters. However, this linear dynamic model has a particular error term structure, combining two terms of unobserved individual heterogeneity. Its estimation requires a new GMM method based on a Double-Difference (two-step) procedure combining first-difference and quasi-difference usual transformations. This estimator is consistent in all cases and can be used to test for the correct specification of the error term following Hausman's approach.

In our sample, the residential water demand is more elastic in the long run (−0.40) than in the short run (−0.26). This result can be explained by the slow adaptation of the households' behaviour to the variation in prices. Price of water has been very low for a long time and consumers could have got into habits of wasting water. Examples of water-using consumer habits are showering, car washing and lawn watering. This result emphasizes the importance of estimating long-run price elasticitics of residential water demand instead of short-run ones. Samples covering longer time periods could also be informative about the possible replacement of durable equipment by some less water consuming. Examples of water-using consumer durables in the home are clothes washers, dishwashers, bathtubs and showers, while those outside include swimming pools, lawns and yard plantings. As a conclusion, long-run price elasticities should be preferred by local authorities when their objective is to assess the impact of tariff policies on the level of consumption and/or on consumer welfare.

Notes

[1] In 1993, 85% of the French population was supplied by a private operator (Guellec, 1995).

[2] The relation between the local public manager and the private operator is frequently described as a power struggle unfavorable to the local community mainly because the elected representatives lack some technical know-how. They are rarely informed on technologies and consequently on costs in water engineering. The local communities could resort to external audits or hire people with technical skills but these solutions are oft en too expensive for small (e.g. rural) communities.

[3] As P_{i0} and $a_s (s = 1, \ldots, t)$ are unknown, we will treat v_i as an error term and θ_t as a temporal parameter in the econometric section. P_{i0} here is a stochastic variable, randomly distributed across municipalities. This stochastic assumption does not affect estimation results in the application below, as the GMM method is used, see Hsiao (1986).

[4] This elasticity measures the relative variation in consumption following a permanent variation in prices.

References

Agthe, D. and Billings, R. (1980) Dynamic models of residential water demand, *Water Resources Research* **16**(3), 476–480.

Ahn, S.C. and Schmidt, P. (1995) Efficient estimation of models for dynamic panel data, *Journal of Econometrics* **68**, 5–27.

Ahn, S.C. and Schmidt, P. (1997) Efficient estimation of dynamic panel data models: Alternative assumptions and simplified estimation, *Journal of Econometrics* **76**, 309–321.

Anderson, T. and Hsiao, C. (1982) Formulation and estimation of dynamic models using panel data, *Journal of Econometrics* **18**, 47–82.

Arellano, M. and Bond, S. (1991) Some tests of specification for panel data: Monte Carlo evidence and an application to employment equations, *Review of Economic Studies* **58**, 277–297.

Arellano, M. and Bover, O. (1995) Another look at the instrumental variable estimation of error-components models, *Journal of Econometrics* **68**, 29–51.

Baltagi, B.H. (1995), *Econometric Analysis of Panel Data*, John Wiley and Sons, New York.

Blundell, R. and Bond, S. (1998) Initial conditions and moment restrictions in dynamic panel data models, *Journal of Econometrics* **87**, 115–143.

Carpenter, A. and Weaver, R. (1997) Damage control productivity: Why econometrics matters, *American Journal of Agricultural Economics* **79**, 47–61.

Carver, P. and Boland, J. (1980) Short- and long-run effects of price on municipal water use, *Water Resources Research* **16**(4), 609–616.

Chichoine, D. and Ramamurthy, G. (1986) Evidence on the specification of price in the study of domestic water demand, *Land Economics* **62**(1), 26–32.

CREDOC (1997) L'eau et les usages domestiques. Comportements de consommation de l'eau dans les ménages, Cahier de Recherche No. 104.

Foster, J.H. and Beattie, B. (1979) Urban residential demand for water in the United States, *Land Economics* **55**(1), 43–58.

Guellec, A. (1995) Le prix de l'eau: de l'exposition à la maîtrise?, Rapport d'information No. 2342, Assemblée National.

Hansen, L. (1982) Large sample properties of generalized method of moment estimators, *Econometrica* **50**, 1029–1054.

Hansen, L. (1996) Water and energy price impacts on residential water demand in Copenhagen, *Land Economics* **72**(1), 66–79.

Hausman, J. (1978) Specification tests in econometrics, *Econometrica* **46**(6), 1251–1271.

Hewitt, J. and Hanemann, W. (1995) A discrete/continuous choice approach to residential water demand under block rate pricing, *Land Economics* **71**(2), 173–192.

Höglund, L. (1997) Estimation of household demand for water in Sweden and its implications for a potential tax on water use, mimeo, University of Göteborg.

Holtz-Eakin, D., Newey, W. and Rosen, H. (1988) Estimating vector autoregressions with panel data, *Econometrica* **56**, 1371–1395.

Howe, C. and Linaweaver, F. (1967) The impact of price on residential water demand and its relation to system design and price structure, *Water Resources Research* **3**(1), 13–32.

Hsiao, C. (1986) *Analysis of Panel Data*, Cambridge University Press, Cambridge.

Mátyás, L. and Sevestre, P. (1992) *The Econometrics of Panel Data – Handbook of Theory and Applications*, Kluwer Academic Publishers, Dordrecht.

Nauges, C. and Thomas, A. (1999) Consistent estimation of dynamic panel data models with time-varying individual effects, mimeo, University of Toulouse I.

Nauges, C. and Thomas, A. (2000) Privately-operated water utilities, municipal price negotiation, and estimation of residential water demand: The case of France, *Land Economics* **76**(1), 68–85.

Nieswiadomy, M. and Molina, D. (1989) Comparing residential water demand estimates under decreasing and increasing block rates using household data, *Land Economics* **65**(3), 281–289.

Point, P. (1993) Partage de la ressource en eau et demande d'alimentation en eau potable, *Revue Economique* **4**, 849–862.

Rivers, D. and Vuong, Q. (1991) Model selection tests for nonlinear dynamic models, Document de Travail, WP 91-08, INRA, Toulouse.

2.2. *Irrigation Water Demand*

Hedonic Price Analysis and Selectivity Bias: Water Salinity and Demand for Land

PHOEBE KOUNDOURI and PANOS PASHARDES*

1. Introduction

Groundwater scarcity has an important qualitative dimension that further limits the supply of usable water. Groundwater quality may affect the productivity of land as an input in agricultural production. Where this is so, the structure of land rents and prices will reflect these environmentally determined productivity differentials. Hence, by using data on land rent or land value for different properties we can in principle identify the contribution which the attribute in question, fresh groundwater quality, makes to the value of (willingness to pay for) the traded good, land. This identifies an implicit or shadow price for the attribute fresh groundwater quality, which in turn can be interpreted as an estimate of the *in situ* scarcity value of the marginal unit of the environmental resource. Methods commonly used to implement this approach include (i) the hedonic technique pioneered by Griliches (1971) and formalized by Rosen (1974); and (ii) the travel cost valuation methods first proposed by Hotelling (1931), and subsequently developed by Clawson (1959) and Clawson and Knetsch (1966). The relationship between land prices and surface and groundwater access (both in quantity and quality terms) has been studied in the hedonic framework by Miranowski and Hammes (1984), Gardner and Barrows (1985), Ervin and Mill (1985),King and Sinden (1988), Caswell and Zilberman (1986) and Torell et al. (1990). Travel cost techniques employed to measure the welfare effects to changes in water quality of recreational sites include Binkley (1978), Freeman (1979), Caulkins et al. (1986), Smith and Desvousges (1986) and Bockstael et al. (1987).

This chapter considers the case where the quality characteristic of an input influences not only the output value but also the usage of the input itself. It is argued that failure to account for alternative uses of an input can give rise to a sample selection problem resulting in misleading parameter estimates reflecting the shadow prices

* We thank the Government of Cyprus and the European Commission, DG XII, for financial support and the Cyprus Ministry of Agriculture, Natural Resources and the Environment, for helping in the collection of the data used in the empirical analysis. The chapter is based in part on work done by the first author for her Ph.D. thesis at the Faculty of Economics and Politics, University of Cambridge. The responsibility for the analysis and interpretation of the data is, of course, ours.

P. Pashardes et al. (eds.),
Current Issues in the Economics of Water Resource Management, 69–80.

of the quality characteristics of the input in question. The sample selection problem here is analogous to the one considered in travel cost models where the endogeneity of the decision to visit a recreational site is shown to result in estimated demand that exaggerates the consumer surplus associated with the trip (Miller and Hay, 1981; Russell and William, 1982; Hellerstein and Mendelsohn, 1992; Hausman et al., 1992).

Our investigation is motivated by the fact that the decision whether to pay for a particular input or not is endogenous to the price paid, as it is the decision how much input to purchase. This is because certain quality characteristics can be responsible for an observation being included in or excluded from the sample. We demonstrate this argument in a model where land close to the seaside is demanded for use as an input either in agricultural production or in touristic development. In the context of this model this selection decision is investigated together with the hedonic valuation of the quality characteristics of the land parcel to avoid the sample selection bias of the type described above. The proximity to the sea, in particular, decreases the probability of land usage for farming due to salination of groundwater supplies and increases the probability of tourism usage due to attractiveness to tourists. Yet, land parcels closed to the seaside may continue to be used in agriculture (in spite of the poor quality of their underwater supplies) because they are still below the touristic development reservation price.

We investigate empirically how this selectivity problem affects the hedonic valuation of the effect of water salinity on agricultural land. The outcome of this empirical investigation is that hedonic valuation techniques might give rise to misleading conclusions about the effect of an environmental attribute on producers (or consumers) welfare if potential biases from inappropriate sample selection criteria are ignored.

The structure of the chapter is as follows. Section 2 describes the decision environment in which the selectivity problem of interest arises. Section 3 considers a model of producer demand for package factors of production and discusses the behavioral effects of characteristics reflecting quality. Section 4 reports the results obtained from the empirical analysis and Section 5 concludes the chapter.

2. Decision Environment

In coastal fresh groundwater systems seawater intrusion is a common form of quality deterioration of groundwater resources that diminishes the water's usefulness for certain purposes. Figure 1 presents a simplified description of the movement of intruding seawater into an aquifer. Consider a coastal irrigation district. A reference boundary R is defined. This can be the coastal perimeter of the irrigation district, the seaward limit of agricultural activity, or an arbitrarily defined line. The object of R is to provide a point of reference from which to measure the length of intrusion.

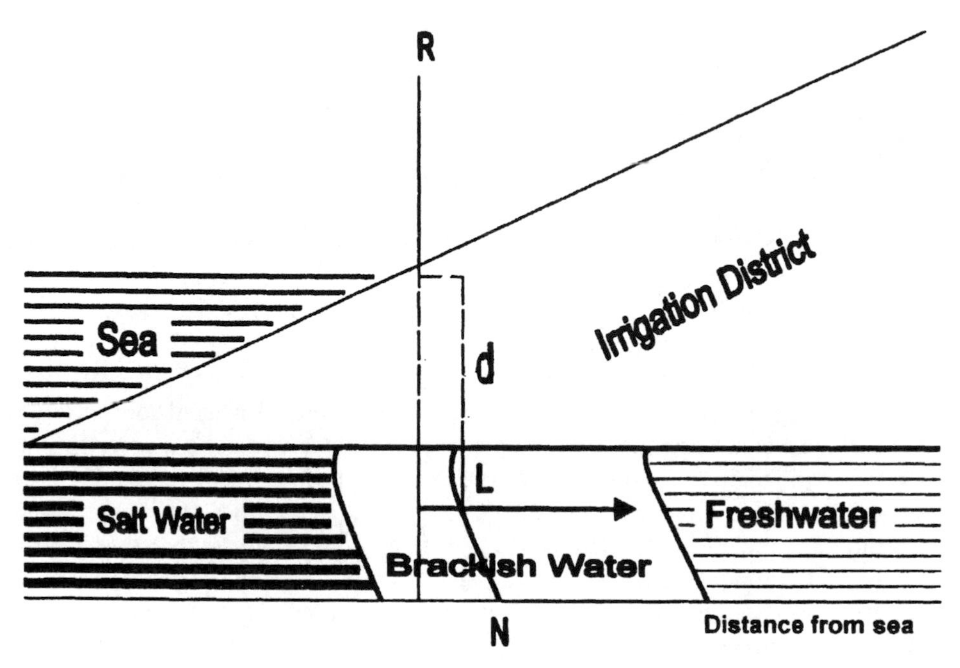

Figure 1. Representation of seawater intrusion.

For an arbitrarily given depth d measured at R, we define the length measure of saltwater intrusion, L. Note that the interface and the point of measuring L is not between saltwater and freshwater. In as much as saltwater and freshwater are miscible fluids, a transition zone will exist between the two fluids. The maximum level of salt concentration for irrigation water is usually around 3000 TDS[1] mg/l. Hence as far as the agricultural sector is concerned interest lies in the interface between water with total salinity greater than 3000 and water with total salinity less than 3000, (the interface between 'brackish water' and 'fresh-water'), as shown in Figure 1.

To understand the movement of saltwater intrusion imagine a district that is divided into n zones. The dimensions of each zone depend on the impact of saltwater intrusion on pumping activity. For example, starting at R, an inland movement of the saltwater interface to a distance N results in the loss of pumping activity in a given area. This arc would be zone 1. The impact of further movements of the interface is usually treated discretely, with each succeeding zone defined in terms of the impact of intrusion on pumping potential. Thus the location of a parcel of land with respect to its proximity to the sea, defines the quality of groundwater supplies accessible to owners of the parcel under consideration: the further away from the sea the parcel is, the lower the impact of seawater intrusion on its groundwater supplies. It is also worth noting that in terms of freshwater stock, instability in the interface between salt water and freshwater causes a widening of the transition (diffusion) zone. Thus as the aquifer is mined, saltwater not only replaces freshwater, but relatively larger quantities of freshwater become brackish. Instability of the interface is directly related to the rate of mining of the aquifer given a level of freshwater stock.

Proximity of a land parcel to the sea can be a proxy for the existence and extent of saltwater intrusion in the parcel's groundwater supplies. As groundwater supplies are contaminated, if other sources for irrigation water are not readily available, new wells must be drilled, coastal injection wells installed, and/or brackish groundwater treated with costly osmosis or catalysis methods. All these imply additional costs for the agricultural producer. Moreover, if salinated water is applied for irrigation, dissolved soils become concentrated in irrigated soils as part of the applied water evaporates through plants and adversely affect crop productivity. Crops vary in their sensitivity to salinity. Generally speaking however, the least sensitive crops are also the least valuable, so areas irrigated with highly saline waters tend to emphasize low-valued types of crops. Thus proximity to the sea decreases the benefits from agricultural use of a parcel of land. As a result, proximity to the sea decreases the probability of land usage for farming. That is, owners of land close to the sea substitute away from fresh groundwater as an input in their production because fresh groundwater becomes more expensive to access due to saltwater intrusion. Hence sea proximity does not only affect the value of agricultural land, but also the decision to use a parcel of land as an input in agricultural production. As shown in the empirical analysis in Section 4, when this endogeneity problem is ignored, a hedonic valuation can understate the effect of groundwater salinization on the price of agricultural land. To correct for this problem we adopt a *Heckman* (1976, 1979) type sample selection process, as described in the next section.

The decision tree relevant for understanding the above argument is graphically presented in Figure 2, where we provide a stylized exposition of the decision process which we model theoretically and estimate empirically in this chapter. Notably, for the construction of this figure, we adopt two of the assumptions that will be employed in the theoretical model of this chapter, to be presented in the following section. Firstly, we assume that the cost function is weakly separable in land, so that prices of other goods can be excluded from the decision to buy a parcel of land (Deaton and Muellbauer, 1980, pp. 122–137). Secondly, we assume that each individual producer purchases only one land bundle.

3. Selectivity and Input Demand

We assume that firm specific production sets over packaged inputs of production are described by the separable cost function

$$C(p, Y_\ell) = G_\ell[c_1(p_{11}, \ldots, p_{1K}), \ldots, c_I(p_{I1}, \ldots, p_{IK}), Y_\ell], \tag{1}$$

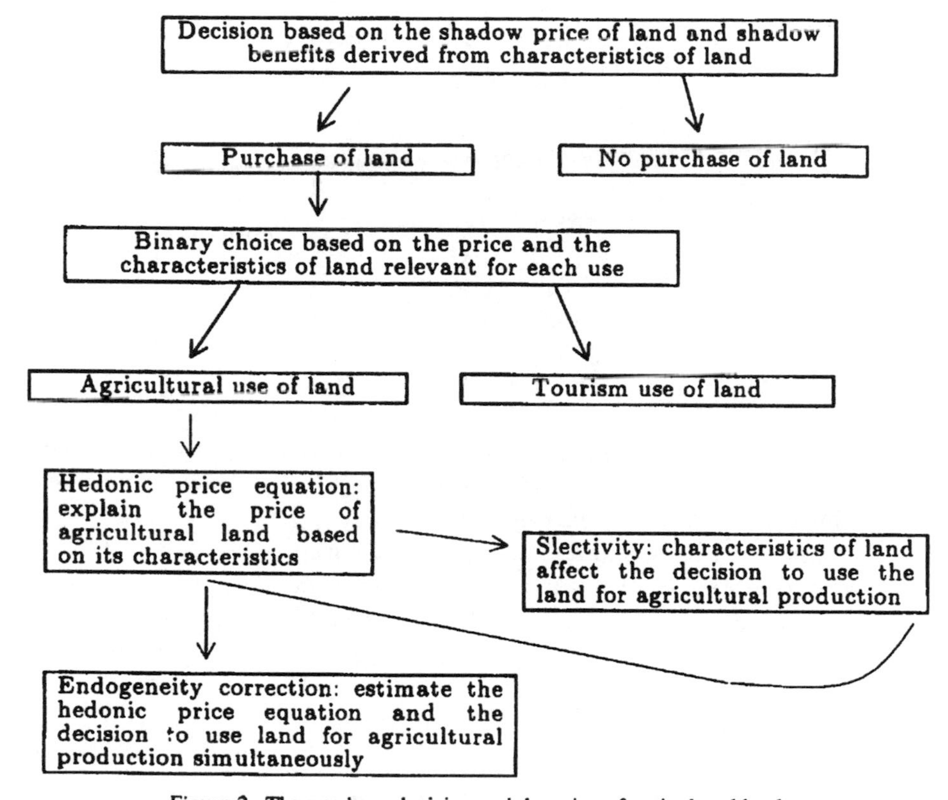

Figure 2. The purchase decision and the price of agricultural land.

where Y_ℓ the units of output produced by the ℓth producer (firm) from the use of all the packaged inputs of production, $c_i(.)$ a sub-function reflecting the unit cost of the ith package and p_{ik} the price of the kth input in this package.

In this context, producer $\ell = 1, \ldots, L$ obtain the $k = 1, \ldots, K$ input indirectly, through purchasing the package $i = 1, \ldots, I$. Applying Shephard's lemma to (1) we obtain demand for the kth input in the ith package by the ℓth producer,

$$q_{ik\ell} = \frac{\partial C(p, Y_\ell)}{\partial p_{ik}} = \frac{\partial G_\ell[.]}{\partial c_i(.)} \frac{\partial c_i(.)}{\partial p_{ik}}, \tag{2}$$

where $\partial G_\ell[.]/\partial c_i(.)$ represents the demand for the ith package and $\partial c_i(.)/\partial p_{ik}$ the *conditional* demand for the individual kth factor of production in this package, i.e. demand for kth input subject to the expenditure of the ith package being decided.

Here we focus on the case where producers purchase only one input package at a time, i.e. the outcome of the above optimization problem is a 'corner solution'. We therefore drop the ith subscript for convenience and incorporate the selection aspect in the analysis by writing expenditure on the selected package by the ℓth producer as

$$\begin{aligned} y_\ell &\equiv \Sigma_k q_{\ell k} p_k = \Sigma_k \left[\frac{\partial G_\ell[.]}{\partial c(.)} \frac{\partial c(.)}{\partial p_k} \right] p_k \\ &= \frac{\partial G_\ell[.]}{\partial c(.)} \Sigma_k \frac{\partial c(.)}{\partial p_k} p_k = \frac{\partial G_\ell[.]}{\partial c(.)} c(.), \end{aligned} \tag{3}$$

where $\partial G_\ell[.]/\partial c(.) = 1$ if the package is selected by the ℓth firm and $\partial G_\ell[.]/\partial c(.) = 0$ otherwise.

We model the package selection decision using a simple Heckman (1976, 1979) type process,

$$I_\ell = g(x_\ell) + v_\ell, \tag{4}$$

where $x_\ell = x_{1\ell}, x_{2\ell}, \ldots, x_{M\ell}$ is a vector of variables affecting the package choice, including quality characteristics and other firm-specific production characteristics like farming skills, etc.; v_ℓ is an error term.

Each firm has a 'reservation' value I_ℓ below which it does not purchase the package in question. As I_ℓ increases (due to quality improvements, increases in firm efficiency, etc.) so does the probability of selecting this package. Using the dummy variable $D_\ell = 1$ when the ith package is selected by the ℓth firm and $D_\ell = 0$ otherwise we can write

$$\begin{cases} D_\ell = 1 \text{ if } g(x_\ell) + v_\ell > I_\ell, \\ D_\ell = 0 \text{ otherwise.} \end{cases} \tag{5}$$

Turning to the modelling of quality heterogeneity of land parcels used by different producers, recall that $c_i(.)$ in (1) is the unit cost of the ith package and define the quality augmented price of the kth input in the package selected by the ℓth producer as $p^*_{k\ell} = \theta_{k\ell} p_k$, where $\theta_{k\ell} \geq 1$. Then quality heterogeneity is introduced in the analysis by writing the unit cost of the package under consideration as

$$c\left(p^*_{1\ell}, \ldots, p^*_{K\ell}\right) = c(\theta_{1\ell} p_1, \ldots, \theta_{K\ell} p_K). \tag{6}$$

At base period prices $p_k = 1$, all k, and (6) obtains the form $c(.) = c(\theta_{1\ell}, \ldots, \theta_{K\ell})$. Then expenditure on the package by the ℓth producer incorporating the selection decision can be written as

$$\ln y_\ell = \ln c(\theta_{1\ell}, \ldots, \vartheta_{K\ell}) + u_\ell \quad \text{if } D_\ell = 1. \tag{7}$$

Truncation from below at $g(x_\ell)$ implies that

$$E[u]_{D_\ell = 1} = \sigma \frac{\phi_\ell(.)}{1 - \Phi_\ell(.)}, \tag{8}$$

where $E[.]$ denotes expectations, $\phi_\ell(.)$ and $\Phi_\ell(.)$ are the values of the probability density and cumulative functions at $g(x_\ell)$, respectively. The simultaneity in the participation and hedonic equations is reflected by $\sigma = \text{cov}(v_\ell, u_\ell)$. If $\sigma = 0$ then selectivity problems are nonexistent, and as a result the hedonic valuation of the effect of water salinity on agricultural land is statistically correct. Therefore, for $D_\ell = 1$ we may write (7) as

$$\ln y_\ell = \ln c(\theta_{1\ell}, \ldots, \theta_{K\ell}) + \sigma M_\ell + \eta_\ell, \tag{9}$$

where M_ℓ is the inverse Mill's ratio[2] and η_ℓ a *random* error term. This equation can be estimated by OLS methods by replacing the unknown M_ℓ values with those computed at $\widehat{g}(x_\ell)$, the predictions obtained from the selection equation (4).

4. Empirical Analysis

The empirical analysis focuses on demand for land parcels by individual agricultural production units. The data are drawn from a Survey of Production (1999) in Kiti, a coastal region located in the island of Cyprus. Data on usage (agriculture or tourism) and price (y_ℓ, $\ell = 1, \ldots, L$) are collected for 193 land parcels. Also collected for these land parcels are many characteristics ($\theta_{k\ell}$, $k = 1, \ldots, K$) reflecting the quality of land, such as the groundwater and soil quality, fragmentation, distance from the sea, distance from the town centre and other environmental and location characteristics.

Property prices, of course, normally capture the marginal value of all possible effects of environmental and location characteristics. As discussed earlier in the chapter, proximity to the sea can give rise to two opposite effects on the price of land used as an input in agricultural production: (a) the probability of land usage for farming decreases and (b) the probability of land usage for tourism development increases. To separate these two effects in our empirical analysis, we use two continuous index variables specifying the proximity of each parcel of land to the coast and the town center of the region under consideration, respectively. These indexes equal one if the parcel is located on the seaside or on the reference point chosen for the town center; and equal zero if the parcel is located at the furthest distance point relevant for the sample of data under consideration.[3]

Assuming that $\ln c(\theta_{1\ell}, \ldots, \theta_{K\ell})$ has the Translog form, we write expenditure on the package by the ℓth producer at base period prices (9) as

$$\ln y_\ell = a_o + \Sigma_k a_k \ln \theta_{k\ell} + \tfrac{1}{2}\Sigma_k \Sigma_j \gamma_{kj} \ln \theta_{j\ell} \ln \theta_{k\ell} + \sigma \widehat{M}_\ell + \eta_\ell, \tag{10}$$

where $\widehat{M}_\ell = \phi_\ell(.)/[1 - \Phi_\ell(.)]$ predicted from a probit equation

$$D_\ell = \lambda_o + \Sigma_s \lambda_s x_{s\ell} + v_\ell, \tag{11}$$

where $D_\ell = 1$ when the ith package is used for farming and $D_\ell = 0$ otherwise and $x_{s\ell}$ includes the $\theta_{k\ell}$ quality characteristics plus years of experience in farming (reflecting farm-specific production skills).

Table 1 reports the parameters obtained from applying (10), the hedonic rental function, to the individual land parcel data described above. The dependent variable is the natural logarithm of the per hectare price of land in Cyprus pounds. Under the heading 'selectivity correction' are the results obtained from the unrestricted version of (10) and under the heading 'no selectivity correction' are the results obtained subject to the restriction $\sigma = 0$.

Commending on the results reported in the first half of Table 1, all variables in the regression model with selectivity correction have effects conforming to expectation. Expenditure on fertilizers, which serves as a proxy for the quality of soil of the parcel, has a significant negative effect on per hectare price of land, apparently indicating the effect of poor soil quality on the selling price of land. Proximity to the town centre has a strongly significant positive effect, a result conforming to the argument that this variable reflects the potential switch of land usage from agricultural production to touristic development. The positive effect of proximity to the town centre on the value of land can also reflect reduction in transportation costs for firms, since the town center is the location where trading of agricultural products takes place.

The variable of interest here, proximity to the coast, does not appear to be significant in the model where a correction for sample selection (farming *versus* tourism) is made by including the Mill's ratio among the explanatory variables. Without this correction however, proximity to the coast appears to have a significant positive effect on the value of agricultural land, apparently indicating that ignoring selectivity correction ignores the fact that the costs of salinization can be offset by an increasing probability of switching to the more lucrative tourism industry. The estimated coefficient indicating the marginal shadow value of groundwater toxicity is not statistically significant at the 95% confidence level in both models (with and without correction for selectivity bias). This result suggests that the cost of increased groundwater toxicity to agriculture cannot be offset by an increasing probability of switching to tourism because, unlike proximity to the sea, this characteristic is not relevant for tourism development.

Regarding the diagnostic tests reported in the lower part of Table 1, it is also worth noting Ramsey's (RESET) specification test suggests correct specification of the estimated model. Also is also worth noting that the LM-test for heteroscedasticity suggests acceptance of the null hypothesis of homoscedastic residuals.

In interpreting the empirical results above recall that in the environmental valuation literature the marginal implicit price of a productive characteristic of land can be derived by differentiating the hedonic value function with respect to that characteristic. In this chapter, the marginal implicit price for coast proximity measures the producer's equilibrium marginal willingness to pay (MWTP) to avoid the marginal increase in the salinization of fresh groundwater supplies beneath her/his land.[4] This MWTP for avoiding coast proximity is estimated to be equal to £10.7 per

Table 1. Parameter estimates of demand for land.

Variable	No selectivity correction		Selectivity correction	
	Parameter	t-ratio	Parameter	t-ratio
Intercept	9.116	32.247	8.233	20.584
Area (0.1 hectares)	-0.686E-02	-2.958	-0.638E-02	-2.803
Existence of House (dummy)	0.495	1.309	0.504	1.364
Existence of Well (dummy)	0.164	1.110	0.131	0.904
Expenditure on Fertilizers (Θ)	-0.469E-03	-2.014	-0.478E-03	-2.099
Value of Investment* (Θ)	-0.134E-04	0.891	0.162E-04	1.097
Proximity to Town (km)	0.224E-02	6.867	0.230E-02	7.200
Groundwater Extraction (m^3)	0.316E-06	1.050	0.390E-06	1.338
Groundwater Toxicity (dummy)	-0.257	-1.641	-0.227	-1.479
Proximity to Coast (km)	0.142	4.316	-0.066	-1.634
Mill's ratio	-	-	3.004	3.053
Diagnostic statistics (model with selectivity correction):				
Number of observations	193			
Mean of dependent variable	7.767			
Adjusted R-squared	0.457			
Ramsey's RESET test	0.876 [F-critical = 1.25]			
LM heteroskedasticity test	0.168 [x^2-critical = 53.7]			

* NPV of construction works and other investments, excluding machinery

hectare of land in the model with selectivity correction whereas the model without selectivity correction indicates that the agricultural producer is willing to pay £11.5 per hectare of land *to gain* a marginal increase in groundwater salinization. Of course, the latter result indicating positive willingness to pay by the average farmer in order to 'gain' marginal degradation in one of the environmental attributes of his land, is counter intuitive. However, it can be explained by the fact that the model without selectivity correction estimates the MWTP for sea proximity which increases the value of land for prospective touristic uses.

5. Conclusion

The argument put forward in this chapter is that hedonic valuations can be misleading when the quality characteristics intended for this valuation have sample selection implications. We consider this argument in the case of land close to the seaside that can be used either as an input in agricultural production or for touristic development. In this case, proximity to the sea can reduce the quality of land as an input in agricultural production, due to salinization of groundwater supplies, but increases the probability of switching the land usage from agriculture to the lucrative tourism market. Deterioration of the groundwater supplies can then appear to have a positive effect on the price of agricultural land.

In the empirical analysis of the chapter the quantifiable water quality attribute is groundwater salinity. Salinity is *ceteris paribus* increasing with sea proximity, the latter being an attribute itself valued in tourist development but not agricultural production. This is the source for sample selectivity bias. The empirical econometric analysis, based on data collected from surveying 282 owners of land parcels, uses Heckman's two step estimation procedure and validates the hypothesis that failing to correct for sample selection results in a biased valuation of groundwater salinity as an attribute of agricultural land. The estimated marginal WTP by the average farmer for fresh groundwater derived from the econometric estimation when correcting for sample selection bias, is statistically not different from zero, whereas without this correction this value appears to have a significant positive effect on the value of land. This result indicates that ignoring selectivity correction ignores the fact that the cost of lower groundwater quality can be offset by an increasing probability of switching to the more lucrative tourism industry. Furthermore, it shows that in a model where the selection and hedonic valuation aspects of agricultural land are modelled simultaneously, low quality groundwater supplies do not have a positive effect on the price of agricultural land.

The overall conclusion of this chapter is that researchers and policy makers in environmental valuation must be careful when employing hedonic techniques to derive willingness to pay for environmental and/or resource quality; it is possible for these techniques to give rise to misleading conclusions about the effect of an environmental attribute on producers (or consumers) welfare if potential biases from inappropriate sample selection criteria are ignored.

The arguments raised here have implications for hedonic price analysis applied to other goods whose quality characteristics can affect sample selection. For example, the approach followed in this chapter may be used to correct for traffic noise appearing to have a potentially positive valuation effect on residential housing because houses in main roads have a high probability of being converted to business properties.

Notes

[1] Although the effects of particular ions on crop productivity vary, the usual approach is to lump all salinity into a macro measure called 'total dissolved solids' (TDS).

[2] The inverse Mill's ratio is given by $\phi_\ell(.)/[1 - \Phi_\ell(.)]$, where $\phi_\ell(.)$ and $\Phi_\ell(.)$ are defined as in the text. This function also represents the hazard function of the truncated normal distribution. As the truncation point increases so does $\Phi_\ell(.)$ and the greater proportion of the population (in the left-hand tail) is discarded and so the mean rises accordingly.

[3] Given that (i) Kiti (the town of the region under consideration) is located on the coast and (ii) touristic development is traditionally more intense closer to the town centre, the distance from the town centre serves as a proxy for the effect of sea proximity on the decision to switch land usage from agricultural production to touristic development.

[4] This leads us naturally to the question of whether producers' inverse demand functions for factor inputs can be identified from observations of marginal implicit prices and quantities. The answer depends on the circumstance of the case. In the model estimated here the hedonic price function is nonlinear and as a result, different producers selecting different bundles of characteristics will have different marginal implicit prices for groundwater quality.

References

Binkley, C. (1978) *The Recreation Benefits of Water Quality Improvements: An Analysis of Day Trips in an Urban Setting*, US Environmental Protection Agency, Washington, DC.

Bockstael, N.E., Hanemann, W.M. and Strand, I.E. (1987) Measuring the benefits of water quality improvements using recreation demand models, Environmental Protection Agency Cooperative Agreement CR-811043-01-0.

Caswell, M.F. and Zilberman, D. (1986) The effects of well depth and land quality on the choice of irrigation technology, *American Journal of Agricultural Economics* **68**, 798–811.

Caulkins, P.P., Bishop, R.C. and Bouwes, N.W. Sr. (1986) The travel cost model for lake recreation: A comparison of two methods for incorporating site quality and substitution effects, *American Journal of Agricultural Economics* **68**, 291–297.

Clawson, M. (1959) Methods of measuring the demand for and value of outdoor recreation, REF Reprint No. 10, Resources for the Future, Washington, DC.

Clawson, M. and Knetsch, J. (1966) *Economics of Outdoor Recreation*, Johns Hopkins University Press, Baltimore, MD.

Deaton, A. and Muellbauer, J. (1980) *Economics and Consumer Behaviour*, Cambridge University Press, Cambridge.

Ervin, D.E. and Mill, J.W. (1985) Agricultural land markets and soil erosion: Policy relevance and conceptual issues, *American Journal of Agricultural Economics* **67**, 938–942.

Freeman, A.M. (1979) *The Benefits of Environmental Improvement: Theory and Practice*, John Hopkins University Press, Baltimore, MD, for Resources for the Future.

Gardner, K. and Barrows, R. (1985) The impact of soil conservation investment on land prices, *American Journal of Agricultural Economics* **67**, 943–947.

Griliches, Z. (ed.) (1971) *Price Indexes and Quality Change*, Harvard University Press, Cambridge, MA.

Hausman, J.A. (1978) Specification tests in econometrics, *Econometrica* **46**, 1251–1271.

Hausman, J., Leonard, G. and McFadden, D. (1992) A utility-consistent, combined discrete choice and count data model: Assessing recreational use losses due to natural resource damage, Paper presented at the Cambridge Economics Symposium titled 'Contingent Valuation: A Critical Assessment', Washington, DC.

Heckman, J. (1976) The common structure of statistical models of truncation, sample selection and limited dependent variables and a simple estimator for such models, *Annals of Economic and Social Measurement* **5**, 475–492.

Heckman, J. (1979) Sample selection bias as a specification error, *Econometrica* **47**, 153–161.

Hellerstein, D. and Mendelsohn, R. (1992) A theoretical foundation for applying count data models to measure recreation values, Working Paper, Economic Research Service, US Department of Agriculture.

Hotelling, H. (1931) The economics of exhaustible resources, *Journal of Political Economy* **39**, 1937–1975.

King, D.A. and Sinden, J.A. (1988) Influence of soil conservation on farm land values, *Land Economics* **64**, 242–255.

Koundouri, P. (2000) Three approaches to measuring natural resource scarcity: Theory and application to groundwater, Ph.D. Thesis, Faculty of Economics and Politics, University of Cambridge.

Miller, J. and Hay, M. (1981) Determinants of Hunter participation: Duck hunting in the Mississippi Flyway, *American Journal of Agricultural Economics* **63**, 677–684.

Miranowski, J.A. and Hammes, D.B. (1984) Implicit prices of soil characteristics for farmland of Iowa, *American Journal of Agricultural Economics* **66**, 645–649.

Rosen, S. (1974) Hedonic prices and implicit markets: Product differentiation in pure competition, *Journal of Political Economy* **2**, 34–55.

Russell, C. and William, J. (1982) The national recreational fishing benefits of water pollution control, *Journal of Environmental Economics and Management* **9**, 328–354.

Smith, V.K. and Desvousges, W.H. (1986) *Measuring Water Quality Benefits*, Kluwer Academic Publishers, Boston, MA.

Torell, A., Libbin, J. and Miller, M. (1990) The market value of water in the Ogallala Aquifer, *Land Economics* **66**, 163–175.

Vaughan, W.J. and Clifford, S.R. (1982) Valuing a fishing day: An application of a systematic varying parameter model, *Land Economics* **60**, 450–463.

PART 3: WATER EFFICIENCY

3.1. *Measurement of Water Efficiency*

Measuring Irrigation Water Efficiency with a Stochastic Production Frontier: An Application to Greek Out-of-Season Vegetable Cultivation

G. KARAGIANNIS, V. TZOUVELEKAS and A. XEPAPADEAS

1. Introduction

Irrigation water is becoming an increasingly scarce resource for the agricultural sector in many regions and countries. A common ground in past policy schemes was the development of adequate irrigation infrastructure to guarantee the supply of irrigation water as the demand for agricultural products was increasing. However, these expansionary policies have resulted in a massive use of irrigation water at a heavily subsidized cost and physical scarcity. Water scarcity has become an increasing social and economic concern for policy makers and competitive water users. Particularly, agriculture is becoming the sector to which policy makers are pointing out at the core of the water problem.

Considerable effort has been devoted over time to introduce policies aiming at increasing water efficiency based on the assertion that "more can be achieved with less water" through better management (Allan, 1999). Better management usually refers to improvement of allocative and/or irrigation water efficiency. The former is closely related to adequate pricing of water use for agricultural purposes, while the latter depends on the type of irrigation technology, environmental conditions, land characteristics, and on scheduling of water application (Omezzine and Zaibet, 1998). In such a context, improving allocative efficiency has been by far the most important (if not the only) factor for conserving water waste and reducing cost of production. But to our view this tendency to focus almost exclusively into allocative efficiency issues is closely related to the definition of irrigation water efficiency adopted in the relevant engineering literature.

Irrigation water efficiency, as previously defined in the literature (McGuckin et al., 1992; Omezzine and Zaibet, 1998), is given by the ratio of effective water use, i.e., the amount of water actually utilized by crop to the water applied to the crop. Based on this definition, a sprinkler irrigation system could reduce water use and increase irrigation efficiency compared to a furrow system, but at the expense of an increase in capital. On the other hand, drip irrigation could be more efficient in water use than sprinklers depending on land characteristics. In purely engineering

P. Pashardes et al. (eds.),
Current Issues in the Economics of Water Resource Management, 85–101.

terms, it has been found that, for surface irrigation methods, average irrigation water efficiency is about 0.6, whereas drip or sprinkler technologies may increase efficiency up to 0.95.

Irrigation water efficiency, as defined above, is a physical measure of a given irrigation technology, presuming a level of management, and thus it is not directly comparable to technical efficiency, as defined by Farrell (1957), which is a measure of management capability. However, as any other production technology, a sprinkler irrigation system for example could possibly be technically inefficient in Farrell's sense due to insufficient training or know how. More importantly, with improper management, a sprinkler irrigation system might use as much water as a furrow system and thus be technically inefficient compared to the well-managed furrow system (McGuckin et al., 1992).

The objective of this chapter is to propose an alternative measure of irrigation water efficiency based on the concept of input-specific technical efficiency, which contracts with measures previously used in the literature. The proposed measure is a non-radial, input-oriented measure of input-specific technical efficiency. It has an economic rather than an engineering meaning and it is defined as the ratio of the minimum feasible water use to observed water use, conditional on production technology and observed levels of output and other inputs used. It provides information on how much water use could be decreased without altering the output produced and the quantities of other inputs used. This measure explicitly recognizes that each irrigation system could be technically inefficient for several reasons that can be explored through statistical methods.

The proposed methodology is applied to a randomly selected sample of 50 out-of-season vegetable growing farms located in Crete, Greece. A stochastic production frontier approach, based on Battese and Coelli's (1995) inefficiency effect model, is used to obtain farm-specific estimates of technical and irrigation water efficiency. In addition, a second-stage regression approach is used to identify the factors influencing irrigation water efficiency differentials across out-of-season vegetable growing farms.

2. Methodological Framework

2.1. Measuring Irrigation Water Efficiency

Let technology be described by the following stochastic production frontier function:

$$y_i = f(x_i, w_i; a)\exp(\varepsilon_i \equiv v_i - u_i), \tag{1}$$

where $i = 1, 2, \ldots, N$ refers to farms, $y \in R_+$ is the quantity of output produced, $x \in R_+^m$ is a vector of input quantities used, w is irrigation water, and

ε_i is a composed error term consisting of a symmetric and normally distributed error term, v_i, representing those factors that cannot be controlled by farmers (i.e., weather effects), measurement errors and left-out explanatory variables, and an one-sided non-negative error term, $u_i \geq 0$, reflecting the shortfall of farm's output from its production frontier, due to the existence of technical inefficiency. Then, farm-specific estimates of output-oriented technical efficiency are obtained as $\mathrm{TE}_i^O = \exp(-u_i)$ (Kumbhakar and Lovell, 2000), while farm-specific estimates of input-oriented technical efficiency are derived by equating (1) with $y_i = f(\vartheta_i x_i, \vartheta_i w_i; \alpha)\exp(v_i)$ and solving for $\mathrm{TE}_i^I = \vartheta_i$ (Atkinson and Cornwell, 1994; Reinhard et al., 1999). Given strict monotonicity, both measures result in the same ranking but in different magnitude of efficiency scores. TE_i^O is greater, equal, or less than TE_i^I whenever returns to scale are decreasing, constant, or increasing, respectively (Fare and Lovell, 1978).

The above measures of efficiency are incapable of identifying the efficient use of individual inputs. For this reason, the proposed irrigation water efficiency measure is based on the non-radial notion of input-specific technical efficiency (Kopp, 1981). Specifically, it is defined as the ratio of minimum feasible to observed use of irrigation water, conditional on the production technology and the observed levels of outputs and inputs. Thus, irrigation water efficiency is an input-oriented, single-factor measure of technical efficiency defined as:

$$\mathrm{IE}^I = [\min\{\lambda : f(x, \lambda w; a) \geq y\}] \rightarrow (0, 1]. \tag{2}$$

Irrigation water efficiency, as defined in (2), has an input-conserving interpretation, which however cannot be converted into a cost-saving measure due to its non-radial nature (Kopp, 1981).

The proposed measure of irrigation water efficiency is illustrated in Figure 1. Let the ith inefficient farmer producing output Y_0 by using x_1 of all other inputs and w_1 units of irrigation water. Then, $\mathrm{TE}_i^I = OB/OA$ and $\mathrm{IE}_i^I = x_1C/x_1A = w_2/w_1$. The proposed irrigation water efficiency measure determines both the minimum feasible water use (w_2) and the maximum possible reduction in water use ($w_1 - w_2$) that still permits the production of Y_0 units of output with unaltered the usage of all other inputs. On the other hand, according to the TE_i^I measure, the maximum possible reduction in water use, required to make the ith farmer technically efficient, is ($w_1 - w_3$). From Figure 1, it is clear that the former will always be greater than the latter. Consequently, the maximum possible reduction in water use suggested by IE_i^I should be considered as an upper bound (Akridge, 1989).

Conceptually, measurement of IE_i^I requires an estimate for the quantity (w_2), which is not observed. Nevertheless, using $\mathrm{IE}_i^I = w_2/w_1$ it can easily be seen that $w_2 = w_1 \cdot \mathrm{IE}_i^I$. By substituting this into (1) and by noticing that point C in Figure 1 lies on the frontier, i.e., $u_i = 0$, (1) may be rewritten as:

$$y_i = f(x_i, w_i^E; a)\exp(v_i), \tag{3}$$

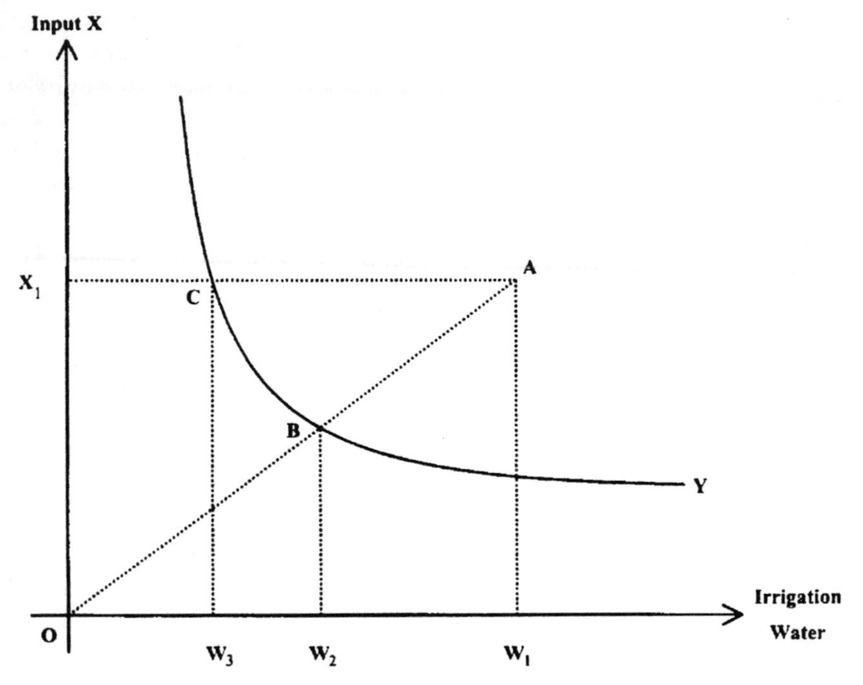

Figure 1. Irrigation water and technical efficiency measures.

where $w_i^E = w_2$ (Reinhard et al., 1999). Then, a measure of IE_i^I can be obtained by equating (1) with (3) and by using the econometrically estimated parameters α.

Since IE_i^I is a non-radial efficiency measure that does not have a direct cost-saving interpretation, the single-factor technical cost efficiency measure could instead be used to evaluate the potential cost savings accruing to more effective management of a single factor (Kopp, 1981). Then, irrigation water technical cost efficiency, ITCE_i, could be defined as the potential cost savings from adjusting irrigation water to a technically efficient level while holding all other inputs at observed levels. Following Akridge (1989), farm-specific estimates of ITCE_i may be obtained as:

$$\mathrm{ITCE}_i = S_{wi}\mathrm{IE}_i^I + \sum_{j=1}^{J} S_{ji}, \tag{4}$$

where S_{wi} and S_{ji} are the ith farm's observed input cost shares for irrigation water and the jth input, respectively.[1] Given that $0 < \mathrm{IE}_i^I \leq 1$ and $S_{wi} + \sum_{j=1}^{J} S_{ji} = 1$ for all i, $0 < \mathrm{ITCE}_i \leq 1$. However, cost savings will vary with factor prices and relatively inefficient water use in a physical sense can be relatively efficient in a cost sense, and *vice versa* (Kopp, 1981).

2.2. Empirical Model

Let the unknown production frontier (1) to be approximated by the following translog specification:

$$\ln y_i = \alpha_0 + \sum_{j=1}^{J} \alpha_j \ln x_{ji} + \frac{1}{2}\left(\sum_{j=1}^{J}\sum_{k=1}^{J} \alpha_{jk} \ln x_{ji} \ln x_{ki}\right) + $$

$$+ \alpha_w \ln w_i + \frac{1}{2}\left(\alpha_{ww} \ln w_i^2 + \sum_{j=1}^{J} \alpha_{jw} \ln x_{ji} \ln w_i\right) + v_i - u_i. \quad (5)$$

Using Battese and Coelli's (1995) inefficiency effect model, the one-sided error term is specified as:

$$u_i = g(z_i; \delta) + \omega_i, \quad (6)$$

where z is a vector of variables used to explain efficiency differentials among farmers, δ is a vector of parameters to be estimated (including an intercept term),[2] and ω_i is an *iid* random variable with zero mean and variance defined by the truncation of the normal distribution such that $\omega_i \geq -[g(z_i; \delta)$. The model (5) and (6) can be estimated econometrically in a single stage using *ML* techniques and the Frontier (version 4.1) computer package developed by Coelli (1992). The variance parameters of the likelihood function are estimated in terms of $\sigma^2 \equiv \sigma_v^2 + \sigma_u^2$ and $\gamma \equiv \sigma_u^2/\sigma^2$, where the γ-parameter has a value between zero and one.

Using the estimated parameters and variances, farm-specific estimates of TE_i^O are obtained as:

$$\mathrm{TE}_i^O = \exp(-u_i) = \exp\left(-\delta_0 - \sum_{m=1}^{M} \delta_m z_{mi} - \omega_i\right). \quad (7)$$

On the other hand, farm-specific estimates of IE_i^I are derived by using (3) and the following relations developed by Reinhard et al. (1999) for the translog specification (5):

$$\mathrm{IE}_i^I = \exp\left[\left\{-\zeta_i \pm \left(\sqrt{\zeta_i^2 - 2\alpha_{ww} u_i}\right)\right\} / \alpha_{ww}\right], \quad (8)$$

where

$$\zeta_i = \frac{\partial \ln y_i}{\partial \ln w_i} = \alpha_w + \sum_{j=1}^{J} \alpha_{jw} \ln x_{ji} + \alpha_{ww} \ln w_i.$$

Given weak monotonicity, a fully technical efficient farm is also irrigation water efficient and thus, only the positive root of (8) is used.[3]

2.3. Explaining Efficiency Differentials

One of the advantages of Battese and Coelli's (1995) model is that it allows measurement of TE_i^O and examination of its differentials among farmers to be done with a single stage estimation procedure. The commonly applied two-stage estimation procedure has been recognized as one that is inconsistent with the assumption of identically distributed inefficiency effects in the stochastic frontier, which is necessary in the *ML* estimation (Reifschneider and Stevenson, 1991; Kumbhakar et al., 1991; Battese and Coelli, 1995).[4] However, the two-stage estimation procedure can be used with no problem for identifying the factors influencing irrigation water efficiency differentials across farms as IE_i^I is calculated from the parameter estimates and the estimated one-sided error component of the stochastic production frontier in (1), and it is not directly related to distributional assumptions. The relevant second-stage regression model has the following form:

$$\ln \mathrm{IE}_i^I = h(z_i, \varepsilon) + e_i, \tag{9}$$

where $h(\cdot)$ is the deterministic kernel of the regression model, ε is the vector of the parameters to be estimated and e_i is an *iid* random variable with zero mean and finite variance.[5] The above model is estimated with *OLS*.

3. Data and Variable Definitions

Data on out-of-season vegetable cultivation were drawn from 50 randomly selected farms operating in Crete, Greece. Questionnaires were collected during the 1998–1999 harvesting period. Farms were located in the four major cultivating regions of Crete, namely Ierapetra, Messara, South Rethymno and Paleochora. To avoid any problems associated with units of measurement, quantity data were converted to indices. For all quantity indices a representative farm were used as a base. Its choice was based on total output and the smallest deviation from sample mean. Aggregation of quantity indices was conducted with Divisia indices. Cost and revenue shares were used to aggregate the quantity indices of aggregate inputs and outputs, respectively.

The dependent variable is the total annual vegetable production measured in Drs. The aggregate inputs considered in the analysis are: (1) total *labor*, comprising of hired (permanent and casual), family and contract labor, measured in working hours; (2) total *seeds* expenses measured in Drs; (3) *chemical inputs*, including fertilizers (nitrogenous, phosphate, potash, complex and others), pesticides and insecticides measured in Drs; (4) *other cost* expenses, consisting of fuel and electric power, irrigation taxes, depreciation expenses,[6] interest payments, fixed assets interest, taxes and other miscellaneous expenses, measured in Drs; and (5) irrigation water measured in m^3.

Table 1. Summary statistics of variables used.

Variable	Mean	St Dev	Min	Max
Output (ths Drs)	5,218	4,642	225	20,900
Labour (working hours)	356	302	61	1,955
Seeds (ths Drs)	403	301	18	3,700
Chemical inputs (ths Drs)	1,228	1,196	60	6,400
Other capital inputs (ths Drs)	1,060	850	155	4,122
Irrigation water (m^3)	146	121	12	700
Extension services (No of visits)	40	25	0	75
Age of farmer (in years)	49	11	27	68
Education (in years)	10	4	6	16
Off-farm income (ths Drs)	1,269	1,141	0	12,000
Bank loans (ths Drs)	3,352	10,808	0	30,000
Share of rented land	0.04	0.14	0.00	0.56
Gross returns (ths Drs per stremma)	2,061	1,855	600	10,500
Farm size (stremmas)	4	3	1	16
Share of hired labor (%)	0.2	0.1	0.0	0.8
Chemical inputs per stremma (ths Drs)	451	368	30	1,620
Participation in local cooperative (%)	72			
Technological innovation[1]	34			
Modern technology[2]	12			
Water source[3]	14			
Sample size (No. of farms)	50			
Ierapetra[4]	29			
Messara[4]	10			
South Rethymno[4]	5			
Paleochora[4]	6			

[1]Percentage of farms adopted technological innovations during the last three years.
[2]Percentage of farms using modern technologies.
[3]Percentage of farms having their own well.
[4]Number of farms located in each of the four regions of Crete.

The explanatory variables in the inefficiency effects (6) and in the second-stage regression (9) include:[7] (1) a dummy variable indicating farms participating in the local agricultural co-operatives; (2) a dummy variable for farms that introduced any technological innovations during the last three years; (3) a dummy variable indicating farms using modern greenhouse technology (some farms in the sample were using a central control system monitoring the greenhouse); (4) a dummy variable indicating farms using water from their own well to irrigate their plantation; (5) the total number of extension visits in the farm; (6) the age of the farmer in years (also age-squared was included in order to examine the possibility of decreasing returns to human capital); (7) the number of years of schooling; (8) the total off-farm income in Drs; (9) yields expressed as total farm output per stremma (one stremma equals 0.1 ha); (10) the total bank loans of the farm; (11) the share of rented land as an indication of farm's tenancy status; (12) the farm's gross returns in Drs per stremma; (13) farm size in stremmas; (14) farming intensity measured as the total value of chemical inputs (fertilizers and pesticides) per stremma; and (15) four regional dummies indicating farms operating in any of the four regions of Crete.

Table 1 presents a summary statistics of variables used. Most of the surveyed farms were selected from the regions of Ierapetra and Messara as these two regions account for the 75% of total out-of-season vegetable production in the island. Out-of-season vegetable cultivating farms are small with an average farm size of 4 stremmas. Only a small portion of that land is rented (0.04%). On the average, gross returns per stremma are 2,061 ths Drs ranging from a low of 600 to a high of 10,500 ths Drs. Annual off-farm income reached on the average 1,269 ths Drs mainly due to tourism, which is the main source off-farm employment. Farmers in the sample seems to be more educated than their counterparts in the other agricultural sectors of the island. The corresponding average figure is 10 years of education compared to 6 years for the whole Cretan agricultural sector. It is also important to note that some farmers (12%) had a higher education degree. The average farmer's age is 49 years ranging from 27 to 68 years. Farmers are enjoying significant extension services (40 visits on average per year) and 72% of them are participating in the local agricultural co-operatives.

A significant part of surveyed farms have introduced some kind of technological innovations during the last three years. These included new genetic improved seeds, irrigation techniques, cultivation practices, and mechanical equipment. Moreover, 12% of the farms are using modern technology such as electronic automated central systems that control the temperature and the moisture of the greenhouse as well as the fertilization, irrigation and chemical spraying of the plantation. However, it seems that the introduction of this modern technology has been associated with large bank loans which are on the average 3,352 ths Drs per farm. Finally, 14% of the farmers are irrigating their plantation using water from their own wells. All of them are located in the greater Ierapetra area were water availability is scarce.

Table 2. Parameter estimates of the translog stochastic production frontier.

Parameter	Estimate	Std Error	Parameter	Estimate	Std Error
α_0	−0.734	(0.042)			
α_L	0.423	(0.127)	α_{SC}	0.000	(0.804)
α_S	0.224	(0.048)	α_{SO}	−0.027	(0.219)
α_C	0.111	(0.028)	α_{SW}	0.000	(0.059)
α_O	0.370	(0.105)	α_{SS}	−0.201	(0.055)
α_W	0.059	(0.011)	α_{CO}	0.000	(0.711)
α_{LS}	0.103	(0.080)	α_{CW}	−0.079	(0.030)
α_{LC}	0.000	(0.433)	α_{CC}	−0.001	(0.001)
α_{LO}	0.309	(0.104)	α_{OW}	0.066	(0.089)
α_{LW}	0.000	(0.521)	α_{OO}	−0.014	(0.006)
α_{LL}	−0.251	(0.116)	α_{WW}	−0.162	(0.048)
σ_2	0.422	(0.045)	γ	0.745	(0.065)
$\ln(\theta)$	−38.756				

Note: L: stands for labour, S: for seeds, C: for chemical inputs, O: for other capital inputs and W: for irrigation water.

4. Empirical Results

4.1. Production Structure

The estimated parameters of the translog stochastic production frontier are presented in Table 2. All the first-order parameters (α_j) have the anticipated (positive) sign and magnitude (being between zero and one), and the bordered Hessian matrix is negative semi-definite indicating that all regularity conditions (i.e., positive and diminishing marginal products) are valid at the point of approximation. On the other hand, the ratio of farm-specific to total variability, γ, is positive and significant at the 1% level. Specifically, the variance of the one-sided component is 0.745, indicating that output-oriented technical efficiency is important in explaining the total variability of output produced. The remaining portion is due to factors outside the control of the farmer (weather, diseases, etc.).

Several hypotheses concerning model specifications are presented in Table 3.[8] From this table it is evident that the traditional average production function does not represent adequately the production structure of out-of-season vegetable cultivating farms in the sample as the null hypothesis that $\gamma = 0$ is rejected at 5% level of significance. Thus, the technical inefficiency effects are in fact stochastic and a significant part of output variability is explained by the existing differences in the degree of output-oriented technical inefficiency. In addition, the hypothesis that the inefficiency effects are absent (i.e., $\gamma = \delta_0 = \delta_m = 0$) is also rejected at 5% level of significance. This indicates that the majority of farms in the sample operate below

Table 3. Model specification tests.

Hypothesis	λ-statistic	Critical value ($\alpha = 0.05$)
$H_0 : \gamma = 0$	29.2	$\chi_2^2 = 5.13^*$
$H_0 : \gamma = 0 = \delta_0 = \delta_m = 0, \forall m$	54.6	$\chi_{20}^2 = 30.8^*$
$H_0 : \delta_0 = \delta_m = 0, \forall m$	51.3	$\chi_{19}^2 = 29.5$
$H_0 : \delta_m = 0, \forall m$	48.9	$\chi_{18}^2 = 28.3$

Note: Critical values with an asterisk are taken from Kodde and Palm (1986, table 1).

Table 4. Production elasticities and returns to scale of out-of-season vegetable cultivating farms in Crete, Greece.

	Average	Ierapetra	Messara	S. Rethymno	Paleochora
Production elasticities					
Labor	0.149	0.269	0.130	0.154	0.164
Seeds	0.505	0.426	0.646	0.507	0.560
Chemicals	0.103	0.076	0.157	0.089	0.110
Other capital	0.316	0.354	0.201	0.255	0.134
Irrigation water	0.053	0.088	0.069	0.048	0.043
Returns to scale	1.127	1.213	1.203	1.062	1.022

the output-oriented technically efficient frontier. Finally, our model specification cannot be reduced neither to Aigner et al. (1977) nor to Stevenson's (1980) models as the null hypothesis of $\delta_0 = \delta_m = 0, \forall m$ and $\delta_m = 0, \forall m$ are rejected at 5% level of significance.

Average estimates of production elasticities and returns to scale are presented in Table 4 for each region under consideration. The estimated production elasticities of all five inputs are positive for all observations. They indicate that in all regions seeds are the foremost important input followed by other capital, labor and chemicals, while irrigation water has the lowest point estimate, which on average were found to 0.053. The latter means that holding all other inputs constant, a 1% reduction in irrigation water requires a sacrifice of 0.53% of marketable output. The highest irrigation water elasticity values were found in Ierapetra and Messara (0.088 and 0.069) and the lowest in South Rethymno and Paleochora (0.048 and 0.043). On the other hand, the hypothesis of constant returns to scale is rejected at the 5% level of significance, and returns to scale were found to be increasing (1.127) with low deviation across regions.

Table 5. Frequency distribution of efficiency ratings of out-of-season vegetable cultivating farms in Crete, Greece.

Efficiency (%)	IE^I	TE^O	ITCE
< 20	0	0	0
20–30	10	0	0
30–40	10	1	0
40–50	13	4	0
50–60	5	2	0
60–70	4	12	0
70–80	5	23	1
80–90	2	8	13
90–100	1	0	36
N	50	50	50
Mean	47.20	70.17	92.49
Minimum	23.09	36.30	76.06
Maximum	98.63	89.09	99.55
Ierapetra	54.18	73.35	93.67
Messara	48.31	66.48	91.90
South Rethymno	33.20	63.75	90.31
Paleochora	35.85	66.30	89.61

A shadow price of irrigation water may be computed by using the mean values of the relevant variables reported in Table 1 and the estimated production elasticity of irrigation water. By combining these figures we find that a reduction of 1.46 m^3 of irrigation water would 'cost' approximately 27.6 ths Drs in terms of forgone revenue. This in turn implies that the shadow price of irrigation water is equal to 18.9 ths Drs per m^3, a value that is much higher than the market price charged in Crete's areas under consideration, which varies between 11 and 15 ths Drs per m^3. This shadow price should be considered as the upper bound of the true shadow price and is only valid for very small change in irrigation water. This is due to the assumption that all other inputs are held constant at their observed levels, which might not be palatable for greater changes in the quantity of irrigation water.[9]

4.2. Technical and Irrigation Water Efficiency

Average estimates of TE_i^O, IE_i^I and $ITCE_i$ are presented in Table 5 in the form of frequency distribution within a decile range. The estimated mean output-oriented technical efficiency is found to be 70.2%, which means that 29.8% increase in production is possible with the present state of technology and unchanged input uses, if technical inefficiency is completely removed. Thus, improving technical efficiency will result to significant increases in farms' revenue and profit. Output-oriented technical efficiency varies from 36.3% to 89.1 and most farms in the sample (62%) achieved technical efficiency greater than 70%. Farms located in Ierapetra and Messara regions exhibited greater efficiency.

Mean irrigation water efficiency is found to be 47.20%, which is much lower than technical efficiency and it also exhibits greater variability ranging from 23.09% to 98.63% (see Table 5). The estimated mean irrigation water efficiency implies that the observed quantity of marketable out-of-season vegetables could have been maintained by using the observed values of other inputs while using 52.8% less of irrigation water. This means that farmers can achieve significant savings in water use if know-how of the utilized irrigation system is improved. It has been argued elsewhere (i.e., Allan, 1999) that water use can be more effective by utilizing more advanced irrigation technologies (e.g., drip irrigation instead of water spreading). It is however evident from the present analysis that besides some farms are using modern greenhouse technologies they do not exploit fully their potential. And unless the potential of a given technology is adequately exploited benefits from that technology may not be as expected.

The concordance between output-oriented technical efficiency and irrigation water efficiency ranking is negative but not strong. The Spearman ranking correlation coefficient between irrigation water and technical efficiency is 0.245. Farms with the highest and lowest irrigation efficiency scores tend generally to exhibit relatively low and high technical efficiency scores, respectively although there are some exceptions. These exceptions refers to farms using modern greenhouse technologies (automated monitoring system). But the majority of relatively technically efficient farms are also water wasting and *vice versa*.

Nevertheless, cost savings that could be attained by adjusting irrigation water to its efficient level would be small since its outlays constitute a small proportion of total cost. For this reason, the estimated mean ITCE_i is much higher than IE_i^I (see Table 5). Specifically, the estimated mean irrigation water technical cost efficiency is found to be 92.5% indicating a potential decrease of 7.5% in total cost by adjusting irrigation water to its efficient level. In addition, the vast majority of farms have achieved irrigation water technical cost efficiency greater than 90%. Thus, even though irrigation water is used least efficiently in technical sense, it offers only few potential cost savings if it is adjusted to its technically efficient level.

Empirical findings concerning the sources of efficiency differentials among farms are presented in Table 6, which contains the results of the inefficiency effects model (6) (second column) and the results of the second-stage regression (9) (third column). With respect to the sign of the estimated parameters, it should be noticed that in the former a negative sign indicates a positive relationship between technical efficiency and the variable under consideration, while in the latter a positive sign

Table 6. Explaining efficiency differentials.

Parameter	TE^O		IE^I	
	Estimate	Std Error	Estimate	Std Error
δ_0	−5.344	(0.654)	4.321	(0.143)
δ_{COP}	0.003	(0.087)	−0.123	(0.233)
δ_{INV}	−0.059	(0.032)	0.204	(0.143)
δ_{MOD}	−0.023	(0.011)	0.005	(0.002)
δ_{WSC}	0.003	(0.098)	−0.243	(0.123)
δ_{EXT}	−0.145	(0.046)	0.213	(0.102)
δ_{AGE}	−0.065	(0.023)	0.102	(0.132)
δ_{AGE2}	0.001	(0.000)	0.045	(0.089)
δ_{EDC}	−0.432	(0.135)	0.089	(0.042)
δ_{OFF}	0.104	(0.044)	−0.123	(0.132)
δ_{INT}	−0.057	(0.021)	−0.084	(0.039)
δ_{CDT}	−0.067	(0.109)	0.092	(0.154)
δ_{TNY}	0.105	(0.054)	−0.155	(0.073)
δ_{GRR}	−0.166	(0.116)	−0.278	(0.302)
δ_{SIZ}	0.034	(0.065)	−0.023	(0.054)
δ_{CHE}	0.089	(0.099)	−0.093	(0.023)
δ_{RG1}	−0.065	(0.021)	0.143	(0.064)
δ_{RG2}	−0.006	(0.003)	0.223	(0.198)
δ_{RG3}	0.076	(0.091)	0.089	(0.121)
δ_{RG4}	−0.089	(0.144)	0.123	(0.143)
R_2			0.415	

Note: COP: is a dummy variable indicating farmers participating in local agricultural co-operative, INV: is a dummy variable indicating farms introduced technological innovations during the last three years, MOD: is a dummy variable indicating farms using modern greenhouse technology, WSC: is a dummy variable indicating the source of irrigation water, EXT: is the number of extension visits, AGE and AGE2: is the farmer's age and age squared in years, EDC: is the number of years of schooling, OFF: is the total off-farm income, INT: is farm's total output per stremma; CDT: is the total bank loans, TNY: is the share of rented land, GRR: is the farm's gross returns per stremma, SIZ: is the farm's size in stremmas, CHE: is the total value of chemical inputs per stremma, RG1, RG2, RG3 and RG4 are regional dummy variables indicating farms located in the four major regions of Crete.

depicts a positive relationship between irrigation water efficiency and the corresponding variable. According to our empirical results, farmers' participation in local co-operatives, bank loans, gross returns per unit of land, and farm size does not seem to affect either technical or irrigation water efficiency. On the other hand, technological innovations, modern greenhouse technology, extension services, and education tend (as expected) to affect positively the degree of both technical and irrigation water efficiency. In contrast, high share of leased land affects negatively both technical and irrigation water efficiency. This most probably is due agency cost between land owners and tenants, reflecting the combined effects of asymmetric information and misaligned incentives between the contracting parties.

The source of irrigation water and intensification (i.e., chemical use per unit of land) affect negatively irrigation water efficiency but they do not affect technical efficiency. It seems that farmers irrigating their plantations using water from their own wells are considerably less irrigation water efficient than those using public domain water, while the negative relationship between intensification and irrigation water efficiency may explain by their strong complementarity in production. On the other hand, farmer's age and off-income affect the degree of technical efficiency but not that of irrigation water efficiency. In particular, it seems that young farmers are becoming relatively more technically efficient over time by improving learning-by-doing, but this would continue until the relationship leveled off and it is expected to decline as farmer approaches the retirement age (Tauer, 1995). In contrast, the degree of technical efficiency tends to decrease as off-farm income increase. Finally, it should be noticed that yields affect positively the degree of technical efficiency, but negatively the efficient use of irrigation water

5. Concluding Remarks

This chapter proposes an alternative measure of irrigation water efficiency based on the concept of input-specific technical efficiency, which contracts with previous physical measures used in the literature. The proposed measure has a pure economic rather than an engineering meaning as it concerns with the managerial capability of farmers rather than the water-saving potentials of each irrigation system. It provides information on how much water use could be decreased without altering the output produced, the technology (including irrigation technology) utilized, and the quantities of other inputs used. The proposed measure explicitly recognizes that each irrigation system could be used inefficiently for several reasons that can be explored through statistical methods. The proposed methodology is applied to a randomly selected sample of 50 out-of-season vegetable growing farms located in Crete, Greece.

Empirical results indicate that irrigation water efficiency is on average much lower than technical efficiency, implying that significant reductions in groundwater waste could be achieved if Cretan out-of-season vegetable growers become more efficient in the use of irrigation water, given the present state of technology and inputs use. This finding contradicts previous beliefs that advanced irrigation practices (i.e., drip irrigation) would by definition improve the efficient use of irrigation water avoiding waste of natural resources. As it is evident from our results, even farmers utilizing modern greenhouse technologies are unable to exploit fully their water- saving potentials. In addition, there is no a clear concordance between technical and irrigation water efficiency. Even though farms that rank high in terms of irrigation water efficiency they usually rank low in terms of technical efficiency, there are some notable exceptions mainly of farmers utilizing modern greenhouse technologies (i.e., automated monitoring system).

According to our findings, modern greenhouse technologies, education, and extension are the main factors associated positively with the degree of irrigation water efficiency. On the other hand, irrigation water efficiency is negatively affected by farming intensity, chemical use, and the percentage of rental land. More importantly, farmers using their own wells exhibited a lower irrigation water efficiency than those using common groundwater sources probably because a price is charged to the latter in a continuous base. These findings provide some initial insights for designing short- and long-run water conservation policies in accordance with the principle that 'more (or at least the same) can be achieved with less water' through better management. No doubt that irrigation water pricing is a core issue in this direction, but following our empirical findings, appropriate use of irrigation systems could also be proven to be an equally important issue in implementing a better water management.

Notes

[1] Farm-specific input price data, or at least information on factor outlays for each farm are required to calculate single-factor technical cost efficiency. The latter has been used in the present study to calculate irrigation water technical cost efficiency.

[2] Exclusion of the intercept parameter may result in biased estimates of δ since in such a case the shape of the distribution of the inefficiency effects is being unnecessarily restricted (Battese and Coelli, 1995).

[3] However, the opposite is not necessarily true. That is, a farm that is irrigation water efficient is not necessarily technically efficient. Moreover, there is no an one-to-one correspondence between the degree of technical and irrigation water inefficiency.

[4] The two-stage estimation approach proceeds as follows. The first stage involves the specification and estimation of the stochastic production frontier function and the prediction of technical inefficiency under the assumption of identically distributed one-side error terms. The second stage involves the specification of a regression model for explaining the predicted technical inefficiency, which however contradicts with the assumption of an identically distributed one-side error term in the stochastic frontier.

[5] Since $0 < \mathrm{IE}_i^I \leq 1$, the dependent variable should be transformed prior to estimation if *OLS* is to be used (Kumbhakar and Lovell, 2000, p. 264).

[6] The depreciation rate applied to machinery varied between 10 and 13% depending on the size of the farm. For buildings (including greenhouses), inventories and land the depreciation rate was 10% of the stock value.

[7] All the explanatory variables except the dummy variables were normalized by their sample means prior to the estimation of the model in order to be indifferent to the units of measurement.

[8] All relevant hypotheses were tested using the generalized likelihood-ratio statistic, $\lambda = -2\{\ln L(H_0) - \ln L(H_1)\}$, where $L(H_0)$ and $L(H_1)$ denote the values of the likelihood function under the null (H_0) and the alternative (H_1) hypothesis. The above test-statistic has approximately a χ^2 distribution, except the case where the null hypothesis involves also $\gamma = 0$. Then, the assumptotic distribution of λ is a mixed χ^2 (Coelli, 1995) and the appropriate critical values are obtained from Kodde and Palm (1986, table 1).

[9] Estimation of actual shadow prices that incorporate substitution possibilities among inputs and output quantity changes requires a profit system approach.

References

Aigner, D.J., Lovell, C.A.K. and Schmidt, P. (1977) Formulation and estimation of stochastic frontier production function models, *Journal of Econometrics* **6**, 21–37.

Akridge, J.T. (1989) Measuring productive efficiency in multiple product agribusiness firms: A dual approach, *American Journal of Agricultural Economics* **71**, 116–125.

Allan, T. (1999) Productive efficiency and allocative efficiency: Why better water management may not solve the problem, *Agricultural Water Management* **40**, 71–75.

Atkinson, S.E. and Cornwell, C. (1994) Estimation of output and input technical efficiency using a flexible functional form and panel data, *International Economic Review* **35**, 245–255.

Battese, G.E. and Coelli, T.J. (1995) A model for technical inefficiency effects in a stochastic frontier production function for panel data, *Empirical Economics* **20**, 325–332.

Coelli, T.J. (1992) A computer program for frontier production function estimation: Frontier version 2.0, *Economics Letters* **39**, 29–32.

Coelli, T.J. (1995) A Monte-Carlo analysis of the stochastic frontier production function, *Journal of Productivity Analysis* **6**, 247–268.

Fare, R. and Lovell, C.A.K. (1978) Measuring the technical efficiency of production, *Journal of Economic Theory* **19**, 150–162.

Farrell, M.J. (1957) The measurement of productive efficiency, *Journal of Royal Statistical Society Series A* **120**, 253–281.

Kodde, D.A. and Palm, F.C. (1986) Wald criteria for jointly testing equality and inequality restrictions, *Econometrica* **54**, 1243–1248.

Kopp, R.J. (1981) The measurement of productive efficiency: A reconsideration, *Quarterly Journal of Economics* **96**, 477–503.

Kumbhakar, S.C. and Lovell, C.A.K. (2000) *Stochastic Frontier Analysis*, Cambridge University Press, New York.

Kumbhakar, S.C., Ghosh, S. and McGuckin, J.T. (1991) A generalised production frontier approach for estimating determinants of inefficiency in US dairy farms, *Journal of Business and Economic Statistics* **9**, 279–286.

McGuckin, J.T., Gollehon, N. and Ghosh, S. (1992) Water conservation in irrigated agriculture: A stochastic production frontier model, *Water Resources Research* **28**, 305–312.

Omezzine, A. and Zaibet, I. (1998) Management of modern irrigation systems in Oman: Allocative vs irrigation efficiency, *Agricultural Water Management* **37**, 99–107.

Reifschneider, D. and Stevenson, R. (1991) Systematic departures from the frontier: A framework for the analysis of firm inefficiency, *International Economic Review* **32**, 715–723.

Reinhard, S., Lovell, C.A.K. and Thijssen, G.J. (1999) Econometric estimation of technical and environmental efficiency: An application to Dutch dairy farms, *American Journal of Agricultural Economics* **81**, 44–60.

Stevenson, R.E. (1980) Likelihood functions for generalised stochastic frontier estimation, *Journal of Econometrics* **13**, 58–66.

Tauer, L.W. (1995) Age and farmer productivity, *Review of Agricultural Economics* **17**, 63–69.

3.2. *Efficient Use and Management of Water*

Strategic Behavior and Efficiency in the Common Property Extraction of Groundwater *

SANTIAGO J. RUBIO and BEGOÑA CASINO

1. Introduction

Groundwater has always been regarded as a common property resource where entry is restricted by land ownership and private exploitation is inefficient. In 1980, Gisser and Sánchez presented a first estimation of this inefficiency, comparing socially optimal exploitation with private (competitive) exploitation using a model with linear water demand, average extraction cost independent of the rate of extraction and linearly decreasing with respect to the water table level. They characterized the private exploitation of the aquifer assuming that farmers are *myopic* and choose their rate of extraction to maximize their current profits, whereas optimal exploitation is obtained by maximizing the present value of the stream of aggregate profits. Their results show that if the storage capacity of the aquifer is relatively large, the difference between the two systems is so small that it can be ignored for practical consideration. This result has been called the Gisser–Sánchez rule by Nieswiadomy (1985) and it has and important policy implication: regulation of the resource does not seem justified since the profits it brings are not very high.[1]

After the publication of the paper by Gisser and Sánchez, several empirical works comparing optimal exploitation and competition have been published: see Feinerman and Knapp (1983), Nieswiadomy (1985), Worthington et al. (1985), Kim et al. (1989), Brill and Burness (1994) and Knapp and Olson (1995). The main conclusion suggested by these studies is that when average extraction cost is assumed to decrease linearly with respect to the water table level, percentage differences in present value are small although nominal differences can be significant, even when uncertainty about surface water supply is taken into account, as reported by Knapp and Olson.

In 1989, Negri distinguished two sources of dynamic inefficiency: a pumping *cost externality* and a *strategic externality*, and used the *differential game theory*

* This paper was presented at the Symposium on Water Resource Management: Efficiency, Equity and Policy, Nicosia, Cyprus, September 22–24, 2000. A first version of this chapter has circulated as Working Paper of the Instituto Valenciano de Investigaciones Económicas (Valencian Institute of Economic Research).

P. Pashardes et al. (eds.),
Current Issues in the Economics of Water Resource Management, 105–122.

to analyze the common property extraction of groundwater.[2] He found, comparing open-loop and feedback equilibria, that the open-loop solution captures only the pumping cost externality whereas the feedback solution captures both, the pumping cost externality and the strategic externality, and it exacerbates the inefficient exploitation of the aquifer compared to the open-loop solution. This result establishes that strategic behavior increases the inefficiency generated by the cost externality. His paper presents a drawback: the existence and uniqueness of the feedback solution are assumed. In Provencher and Burt (1993) optimal and feedback equilibria, computed using discrete-time dynamic programming, are also compared. The authors explore dynamic inefficiencies via Kuhn–Tucker conditions. They conclude that if the value function is concave then strategic behavior increases the inefficiency of private groundwater exploitation. Furthermore they assert that the steady-state groundwater reserves attained when firms use decision rules strategies are bounded from below by the steady-state arising when firms are myopic, and from above by the steady-state arising from optimal exploitation.

However, these results are different from those obtained by other authors such as Tsutsui and Mino (1990), in the field of industrial economics, or Dockner and Long (1993), Wirl (1994) and Wirl and Dockner (1995), in the field of environmental economics. Tsutsui and Mino examine, for a differential game of duopolistic competition with sticky prices, whether it is possible to construct a more efficient feedback equilibrium using *nonlinear* strategies. They find that there exist feedback equilibria supported by nonlinear strategies which approach the cooperative solution more than the open-loop equilibrium. Dockner and Long (1993) have obtained identical results for a differential game of international pollution control with two countries and nonlinear strategies, and Wirl (1994) and Wirl and Dockner (1995) have proved that cooperation between an energy cartel and a consumers' government is not necessary to reach the efficient long-run concentration of CO_2 in the atmosphere.

In this chapter we adapt the model defined by Gisser and Sánchez with the aim of examining whether strategic behavior plays *against* the efficiency of the solution, as established by Negri and Provencher and Burt, or *for* the efficiency, as seemed to happen in Tsutsui and Mino's, Dockner and Long's and Wirl's papers. In other words, we are interested in studying whether it is possible, following Tsutsui and Mino's procedure, to construct a more efficient feedback equilibrium using *nonlinear* strategies instead of *linear strategies*. Our results show that the procedure proposed by Tsutsui and Mino is not useful for the differential game studied in this chapter. In fact, because of the *local* nature of the nonlinear strategies, this procedure can only be applied when the domain of state variable initial value is not restricted. However this situation does not occur in our groundwater pumping differential game, as long as the initial value of water table is not an *arbitrary* value but the value corresponding to the (*natural*) *hydrologic steady-state*. As a corollary of this result we establish the existence of a *unique* stable steady-state supported by a Markov perfect feedback equilibrium which is reached when the agents play linear strategies. Finally, our findings show that strategic behavior plays against the efficiency of private exploitation, but they also confirm the applicability of the Gisser–Sánchez rule.[3]

In the next section we present our formulation of the differential game and we derive the open-loop Nash equilibrium and the efficient equilibrium in Section 3. The optimal (linear and nonlinear) Markov strategies are derived in Section 4. A discussion on linear *versus* nonlinear strategies and a comparison of the different steady-states computed in the chapter are presented in Section 5. The chapter ends with some concluding remarks.

2. The Model

We assume that demand for irrigation water is a negatively sloped linear function

$$W = g + kP, \quad k < 0, \tag{1}$$

where W is the pumping and P is the price of water. We also assume that farmers sell their production in competitive markets so that the price of water is equal to the value of its marginal product.[4] Moreover, the production function presents constant returns to scale and production factors, other than water and land, are optimized subject to the rate of water extraction.

Access to the aquifer is restricted by land ownership and consequently the number of farmers is fixed and finite over time. Additionally, we assume that all farmers are identical. Using symmetry we can write the aggregate rate of extraction as $W = Nw_i$, where N is the number of farmers and w_i the rate of extraction of the representative farmer. Thus, the individual demand functions are

$$w_i = \frac{1}{N}(g + kP), \quad i = 1, \ldots, N \tag{2}$$

and the revenues of the ith farmer

$$\int p(w_i)\,\mathrm{d}w_i = \frac{N}{2k}w_i^2 - \frac{g}{k}w_i. \tag{3}$$

The total cost of extraction depends on the quantity of water extracted and the depth of the water table

$$C(H, W) = (c_0 + c_1 H)W, \quad c_1 < 0, \tag{4}$$

where H is the water table elevation above some arbitrary reference level. If this arbitrary reference point is the bottom of the aquifer then c_0 is the maximum average cost of extraction and $H_m = -c_0/c_1$ represents the maximum water table elevation associated with the *hydrologic steady-state* of the aquifer occurring when groundwater reserves reach the storage capacity of the aquifer. Then, as the marginal and average costs do not depend on the rate of extraction, the individual farmer's extraction costs are

$$C_i(H, w_i) = \frac{1}{N}(c_0 + c_1 H)W = (c_0 + c_1 H)w_i. \tag{5}$$

Costs vary directly with the pumping rate and inversely with the level of the water table. Marginal and average costs increase with the pumping lift and are independent of the extraction rate. We are implicitly assuming that changes in the water level are transmitted instantaneously to all users. This assumption clearly exaggerates the degree of common property. Moreover, the symmetry assumption requires the groundwater basin to have parallel sides with a flat bottom.

The differential equation describing the dynamics of the water table is obtained as the difference between natural recharge and net extractions

$$\mathrm{AS}\,\dot{H} = R + (\gamma - 1)W, \quad H(0) = H_m > 0 \tag{6}$$

where R is the natural recharge, γ is the return flow coefficient defined between zero and unity, and AS is the area of the aquifer times storativity. We assume that the rate of recharge is constant and deterministic and, although artificial recharge of the aquifer is feasible, we focus on the case where the resource is being depleted.[5]

Finally, we assume that interactions among agents are completely noncooperative and rational, so the ith farmer faces the following dynamic optimization problem:

$$\max_{\{w_i\}} \int_0^\infty e^{-rt} \left[\frac{N}{2k} w_i^2 - \frac{g}{k} w_i - (c_0 + c_1 H) w_i \right] \mathrm{d}t, \tag{7}$$

where r is the discount rate and the dynamics of H is given by (6). We implicitly assume the nonnegativity constraint on the control variable and we do not impose $H \geq 0$ as a state constraint but as a terminal condition: $\lim_{t\to\infty} H(t) \geq 0$ for simplicity.[6]

3. Open-Loop Nash Equilibrium

In the open-loop Nash equilibrium, farmers commit themselves when starting the game to an entire temporal path of water extraction that maximizes the present value of their stream of profits given the extraction path of rival farmers.[7] Then for every given path $w_j(t)$ of farmer j, $j = 1, \ldots, N-1$, farmer i faces the problem of maximizing (7) given $w_j(t)$. Simultaneously, the other players j face a similar problem. An equilibrium of the game consists of N open-loop strategies that solve the N optimization problems simultaneously. The necessary conditions for an interior open-loop equilibrium are

$$\frac{N}{k}w_i - \frac{g}{k} - (c_0 + c_1 H) + \lambda_i \frac{\gamma - 1}{\mathrm{AS}} = 0, \quad i = 1, \ldots, N, \tag{8}$$

$$\dot{\lambda}_i = r\lambda_i + c_1 w_i, \quad i = 1, \ldots, N, \tag{9}$$

where λ_i is the costate variable or the user cost of the resource.[8]

With symmetry, $w_i = w_j = w$ and $\lambda_i = \lambda_j = \lambda$ and therefore the $2N$ equations defined by (8) and (9) reduce to 2.[9] Taking into account that at the steady-state $\dot{H} = \dot{\lambda} = 0$, Equations (6) and (9) can be used to find the following steady-state values

$$w^* = -\frac{R}{(\gamma - 1)N}, \quad \text{and} \quad \lambda^*_{\mathrm{OL}} = \frac{c_1 R}{r(\gamma - 1)N}. \tag{10}$$

Then by substitution in (8) we obtain the steady-state value of water table

$$H^*_{\mathrm{OL}} = -\frac{R}{kc_1(\gamma - 1)} + \frac{R}{r\,\mathrm{AS}\,N} - \frac{1}{c_1}\left(\frac{g}{k} + c_0\right). \tag{11}$$

The above result can be summarized as follows:

PROPOSITION 1. *There exists a unique steady-state for the open-loop Nash equilibrium of the game. The steady-state values are given by (10) for the rate of extraction and by (11) for the water table.*

Observe that in this game as the dynamics of water table H does not depend on H, the stationary equilibrium extraction rate is independent of the equilibrium concept used to resolve the game.

To evaluate the efficiency of the market equilibrium we need the *socially optimal or efficient* equilibrium. This equilibrium can be obtained as a particular case of the open-loop Nash equilibrium making N equal to the unity.[10]

$$H^*_{\mathrm{SO}} = -\frac{R}{kc_1(\gamma - 1)} + \frac{R}{r\,\mathrm{AS}} - \frac{1}{c_1}\left(\frac{g}{k} + c_0\right). \tag{12}$$

Now we can compare the stationary values of the water table for the two equilibria

$$\Delta_1 H^* = H^*_{\mathrm{SO}} - H^*_{\mathrm{OL}} = \frac{R}{\mathrm{AS}\,r}\left(1 - \frac{1}{N}\right) > 0. \tag{13}$$

This difference represents the impact of the *pumping cost externality* on the stationary value of the water table. These results allow us to present the following proposition:

PROPOSITION 2. *The socially optimal steady-state water table is higher than the steady-state water table supported by the open-loop Nash equilibrium and the difference declines as the discount rate increases or the number of farmers decreases.*

The effect a discount rate variation has on the difference between the two steady-state values is explained by the different impact that a variation in the discount rate has on user cost in each case. As $\lambda^*_{OL} N = \lambda^*_{SO}$ we find that $|\partial\lambda^*_{OL}/\partial r| < |\partial\lambda^*_{SO}/\partial r|$. Thus, an increase in the discount rate means a decrease of the user cost in both cases, but by a larger amount in the optimal solution. Furthermore, the effect a variation in the number of farmers has on the difference is clear, bearing in mind that the socially optimal equilibrium is independent of the number of farmers. Thus an increase in the number of farmers reduces the user cost of the resource and, consequently, the steady-state water table for the open-loop Nash equilibrium, causing an increase in the difference between the two steady-state values.[11]

4. Stationary Markov Feedback Equilibrium

In an economic environment in which binding commitments are not feasible because of undefined property rights and where all players have access to current information on water table elevation, strategies that depend only on time cannot be credible. As is well known, this requirement of credibility is fulfilled by a stationary Markov feedback equilibrium derived by the dynamic programming approach. In Markov feedback equilibria each farmer adopts a decision rule that depends on the water table, which reflects the extraction rates of the other farmers, taking as given the decision rules of their rivals. That is, when farmers play feedback strategies they do not respond directly to the extraction rates of the others; rather, as extraction rates may not be observed, information about the rival's behavior is obtained by observing the stock of water.

In our differential game, linear stationary Markov strategies can be guessed.[12] However, as Tsutsui and Mino (1990) state, feedback equilibria obtained by the guessing method are typically less efficient than the open loop equilibrium and far less efficient than the collusive outcome. Thus, these authors suggest a more general method that allows for other feedback equilibria excluded by the guessing method. For this reason, in this section we use the procedure proposed by Tsutsui and Mino to obtain Markov feedback equilibria. As noted above, a stationary Markov feedback equilibrium must satisfy the Hamilton–Jacobi–Bellman equation of dynamic programming, which for infinite horizon autonomous problems as ours is written as

$$rV_i(H) = \max_{\{w_i\}} \left\{ \left[\frac{N}{2k} w_i^2 - \frac{g}{k} w_i - (c_0 + c_1 H) w_i \right] \right.$$

$$+ V_i'(H)\left[\frac{1}{\text{AS}}\left(R + (\gamma - 1)\sum_{i=1}^{N} w_i\right)\right]\Bigg\}, \tag{14}$$

where $i = 1, \ldots, N$, $V_i(H)$ stands for the optimal current *value function* associated with problem (7); i.e., it denotes the maxima of the objective (7) subject to (6) for the current value of the state variable, and $V_i'(H)$ is its first derivative.

Using the maximization condition

$$V_i'(H) = (\text{AS}/(\gamma - 1))((g/k) + c_0 + c_1 H - (N w_i/k))$$

and the symmetry assumption we have

$$\begin{aligned} rV(H) &= \left[\frac{N}{2k}w^2 - \left(\frac{g}{k} + c_0 + c_1 H\right) w\right] \\ &\quad + \left(\frac{R}{\gamma - 1} + Nw\right)\left(\frac{g}{k} + c_0 + c_1 H - \frac{N}{k}w\right). \end{aligned} \tag{15}$$

Temporarily assuming a zero discount rate, the Hamilton–Jacobi–Bellman equation (15) becomes the quadratic equation in w

$$\begin{aligned} 0 &= -N\left(\frac{2N-1}{2k}\right) w^2 + \left[(N-1)\left(\frac{g}{k} + c_0 + c_1 H\right) - \frac{NR}{k(\gamma - 1)}\right] w \\ &\quad + \frac{R}{\gamma - 1}\left(\frac{g}{k} + c_0 + c_1 H\right) \end{aligned} \tag{16}$$

with two nonlinear solutions

$$\begin{aligned} w_{1,2} &= \frac{k}{N(2N-1)}\Bigg\{(N-1)\left(\frac{g}{k} + c_0 + c_1 H\right) - \frac{NR}{k(\gamma - 1)} \\ &\quad \pm \Bigg[\left((N-1)\left(\frac{g}{k} + c_0 + c_1 H\right) - \frac{NR}{k(\gamma - 1)}\right)^2 \\ &\quad + \frac{2N(2N-1)R}{k(\gamma - 1)}\left(\frac{g}{k} + c_0 + c_1 H\right)\Bigg]^{1/2}\Bigg\}. \end{aligned} \tag{17}$$

Based on this result for $r = 0$ we propose a nonlinear strategy for the case with $r > 0$ given by

$$w(H) = \frac{1}{N(2N-1)}\left[k(N-1)\left(\frac{g}{k} + c_0 + c_1 H\right) - \frac{NR}{(\gamma - 1)}\right] + f(H), \tag{18}$$

where $f(H)$ is a nonlinear function in H. Working with (18) and the Hamilton–Jacobi–Bellman equation we obtain a set of stationary Markov strategies implicitly defined by the equation[13]

$$K = \left\{w - \frac{1}{N(2N-1)}\left[k(N-1)\left(\frac{g}{k} + c_0 + c_1 H\right) - \frac{NR}{(\gamma-1)}\right] - \left(H + \frac{G}{F}\right)y_a\right\}^{\xi_1}$$
$$\times \left\{w - \frac{1}{N(2N-1)}\left[k(N-1)\left(\frac{g}{k} + c_0 + c_1 H\right) - \frac{NR}{(\gamma-1)}\right] - \left(H + \frac{G}{F}\right)y_b\right\}^{\xi_2}, \tag{19}$$

where K is an arbitrary constant, and

$$F = -\frac{k(N-1)^2 c_1^2}{N(2N-1)} + \frac{N\,\mathrm{AS}\,c_1 r}{(\gamma-1)(2N-1)} > 0,$$

$$G = \frac{F}{c_1}\left(\frac{g}{k} + c_0\right) - \frac{Nk(\gamma-1)c_1 - N\,\mathrm{AS}\,r}{k(2N-1)(\gamma-1)^2}R,$$

$$y_{a,b} = \frac{1}{2}\left\{\frac{r\,\mathrm{AS}}{(\gamma-1)(2N-1)} \pm \left[\left(\frac{r\,\mathrm{AS}}{(\gamma-1)(2N-1)}\right)^2 - \frac{4kF}{N(2N-1)}\right]^{1/2}\right\},$$

where

$$\xi_1 = \frac{y_a}{y_b - y_a} < 0 \quad \text{and} \quad \xi_2 = \frac{-y_b}{y_b - y_a} < 0.$$

This function is nonlinear in the control variable except for $K = 0$. In that case, Equation (19) admits two *linear* solutions:

$$w_a = \frac{k(N-1)}{N(2N-1)}\left(\frac{g}{k} + c_0\right) - \frac{R}{(\gamma-1)(2N-1)} + \frac{G}{F}y_a + \left[\frac{k(N-1)c_1}{N(2N-1)} + y_a\right]H, \tag{20}$$

$$w_b = \frac{k(N-1)}{N(2N-1)}\left(\frac{g}{k} + c_0\right) - \frac{R}{(\gamma-1)(2N-1)} + \frac{G}{F}y_b + \left[\frac{k(N-1)c_1}{N(2N-1)} + y_b\right]H. \tag{21}$$

Therefore, the set of solution curves given by (19) consists of two straight lines and a family of hyperbolic curves represented in Figure 1. Notice that each solution

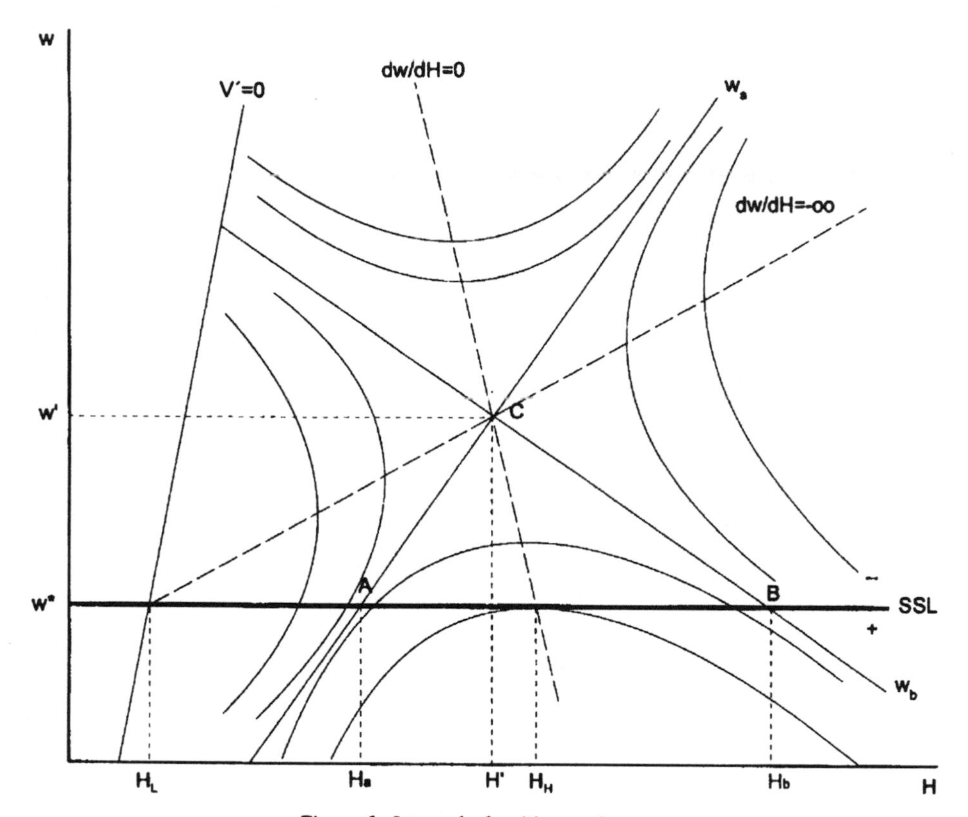

Figure 1. Interval of stable steady-states.

curve is only well defined in the region of nonnegative marginal value, that is, on the right of the line defined by $V' = 0$. Each curve corresponds to a nonlinear stationary Markov feedback equilibrium and the two straight lines, $w_a(H)$ and $w_b(H)$, correspond to the singular solutions (20) and (21). They are positively and negatively sloped, respectively, and go through (H', w'). The steep dotted line is the locus $\mathrm{d}w/\mathrm{d}H = -\infty$, whereas the dotted line with negative slope is the locus $\mathrm{d}w/\mathrm{d}H = 0$, for Equation (19).

As in our game the set of solution curves covers the entire H–w plane, we see that for each point on the steady-state line, SSL, (defined by $w^* = -R/N(\gamma - 1)$) it is always possible to find a strategy that can satisfy the stationary condition: $\dot{H} = 0$. However, the existence of a steady-state value does not necessarily mean that there exists a path, $H^*(t)$, that converges to it. For this reason we are interested in the stable steady-states. Taking into account that the rate of extraction is given by the nonlinear stationary Markov strategies implicitly defined by (19), the state equation (6) can be written as

$$\dot{H} = \frac{1}{\mathrm{AS}}\left[R + (\gamma - 1)Nw(H)\right].$$

Linearizing this equation around the steady-state gives the stability condition: $\mathrm{d}w/\mathrm{d}H > 0$, which implies that a steady-state water table H^* is *locally* stable when the slope of $w(H)$ is positive at the intersection point with SSL. Graphically this means that the set of locally stable steady-states is defined by the interval (H_L, H_H) (see Figure 1). The limits of this interval can be calculated as the intersection points of the lines $V' = 0$ and $\mathrm{d}w/\mathrm{d}H = 0$ with SSL. We find that these two values coincide with the steady-state values of the myopic and open-loop equilibria calculated in Section 3.[14]

$$H_L = H^*_{\mathrm{MY}} = -\frac{R}{kc_1(\gamma - 1)} - \frac{1}{c_1}\left(\frac{g}{k} + c_0\right), \tag{22}$$

$$H_H = H^*_{\mathrm{OL}} = \frac{R}{kc_1(\gamma - 1)} + \frac{R}{r\,\mathrm{AS}\,N} - \frac{1}{c_1}\left(\frac{g}{k} + c_0\right). \tag{23}$$

The next proposition summarizes these results.

PROPOSITION 3. *Any water table level in the interval (H_L, H_H) is a locally stable steady-state, where H_L is the steady-state water table for the myopic solution and H_H is the steady-state water table for the open-loop solution.*

5. Linear *versus* Nonlinear Strategies

As we can see from Figure 1, the steady-state water table in the interval (H_L, H_H) that can be supported by a stationary Markov strategy depends on the initial water table level. If the *initial* water table is higher than H_b, which is defined by the intersection of the unstable linear strategy and the steady-state line, only the stable *linear* strategy leads to a steady-state. However, if the *initial* water table is lower than H_b there exists a multiplicity of steady-states which can be reached using stable *nonlinear* strategies. In this chapter we assume first that the initial water table level, $H(0)$, is equal to its *(natural) hydrologic steady-state*, H_m, corresponding to the maximum water table elevation at which the water reserves coincide with the storage capacity of the aquifer, and second that human activity, justified by economic parameters, consists of *mining* the aquifer until an *economic-hydrologic steady-state* is reached. The difference between the two equilibria is that the first depends exclusively on hydrologic parameters whereas the second depends on hydrologic and economic parameters. This assumption has a clear consequence: the socially optimal steady-state water table, H^*_{SO}, must be lower than the initial water table, H_m. Now if we compare H_b with H^*_{SO} in order to establish the relationship between H_b and H_m, we find that H_b is lower than H^*_{SO}, and, consequently, that H_b is lower than H_m (see Appendix B). Then as the *initial* water table level is higher than H_b, only the stable *linear* strategy leads to a steady-state water table given by H_a, the intersection point between the stable *linear* strategy, $w_a(H)$, and the steady-state line:

$$H_a = H^*_F = -\frac{R}{kc_1(\gamma - 1)} D_a - \frac{1}{c_1}\left(\frac{g}{k} + c_0\right), \tag{24}$$

where

$$D_a = \frac{\frac{(N-1)kc_1}{N(2N-1)} + \frac{N^2[kc_1(\gamma-1) - r\,\mathrm{AS}]}{(N-1)^2 kc_1(\gamma-1) - N^2 r\,\mathrm{AS}} y_a}{\frac{(N-1)kc_1}{N(2N-1)} + y_a}. \tag{25}$$

The above result can be summarized as follows:

PROPOSITION 4. *There exists a unique steady-state supported by a Markov feedback equilibrium in linear strategies. The steady-state water table is given by (24).*

This result limits the scope of the procedure proposed by Tsutsui and Mino (1990) to construct a feedback equilibrium using *nonlinear* strategies. In fact, our conclusion is that their procedure is not useful for the differential game proposed in this chapter. The problem with Tsutsui and Mino's proposal is that, given the *local* nature of the nonlinear strategies, the stable steady-state water tables can only be reached from a restricted domain of water table initial values. For instance, to reach a steady-state water table in the interval (H_a, H_H), the initial water table must be in the interval (H_H, H_b). In other words, Tsutsui and Mino's proposal is useful when the domain of state variable initial value is not restricted. But this does not happen in our groundwater pumping differential game and maybe this does not happen either in another natural resource and environmental models, as long as the initial value of the state variable for these models is not an *arbitrary* value but the value corresponding to the *natural steady-state*.[15]

Returning to the discussion on the evaluation of the strategic externality effect on the steady-state value of the water table, we find that this effect is given by the difference between H^*_{OL} and H^*_F:

$$\Delta_2 H^* = H^*_{OL} - H^*_F = R\frac{(D_a - 1)r\,\mathrm{AS}\,N + kc_1(\gamma - 1)}{kc_1(\gamma - 1)r\,\mathrm{AS}\,N}. \tag{26}$$

One can graphically check that this difference is positive (see Figure 1), so that we can conclude that the strategic externality exacerbates the dynamic inefficiency of groundwater private exploitation generated by the cost externality.[16] Moreover, taking into account the relationship existing among H^*_{MY}, H^*_F and H^*_{OL}, an *upper bound* can be defined for this effect given by

$$\Delta_2 H^* < H^*_{\mathrm{OL}} - H^*_{\mathrm{MY}} = \frac{R}{r\,\mathrm{AS}\,N}. \tag{27}$$

These results are summarized in the last proposition.

PROPOSITION 5. (i) *The steady-state water table supported by the open loop Nash equilibrium is higher than the steady-state water table supported by the Markov feedback equilibrium.* (ii) *The impact of the strategic externality on the steady-state value of the water table presents an upper bound given by the difference between the steady-state water table for the open-loop solution and the steady-state water table for the myopic solution. This difference declines as the discount rate and/or the number of farmers increases.*[17]

Finally, once we have established the relationship between H^*_{OL} and H^*_F, we are able to compare the steady-state values of the water table for the different solutions studied in this chapter

$$H^*_{\mathrm{MY}} < H^*_F < H^*_{\mathrm{OL}} < H^*_{\mathrm{SO}}, \tag{28}$$

where $H^*_{\mathrm{SO}} - H^*_{\mathrm{OL}}$ is the effect of *cost externality*, and the difference $H^*_{\mathrm{OL}} - H^*_F$ is the effect of *strategic externality*. Thus, the inefficiency of the private solution, evaluated as the divergence between the private solution steady-state water table and the socially optimal steady-state water table, is given by $H^*_{\mathrm{SO}} - H^*_F$. Now, we can calculate using (13), the effect of the cost externality, and (26), the effect of strategic externality, the dynamic inefficiency associated with the private exploitation of groundwater.

$$\Delta H^* = \Delta_1 H^* + \Delta_2 H^* = R\frac{(D_a - 1)r\,\mathrm{AS} + kc_1(\gamma - 1)}{kc_1(\gamma - 1)r\,\mathrm{AS}}. \tag{29}$$

This difference presents an *upper bound* given by

$$\Delta H^* = H^*_{\mathrm{SO}} - H^*_F < H^*_{\mathrm{SO}} - H^*_{\mathrm{MY}} = \frac{R}{r\,\mathrm{AS}}. \tag{30}$$

So that we can conclude that the difference between *socially optimal exploitation* and *private exploitation* of a groundwater, characterized by a stationary Markov feedback equilibrium in linear strategies, decreases with the storage capacity of the aquifer and if this is large enough the two equilibria are almost identical. In fact, Gisser and Sánchez obtain in their paper the same expression that appears in the right-hand side of (30).[18] Obviously, this limits the practical scope of Negri's result if the storage capacity is large enough.[19]

6. Conclusions

In this chapter we have developed the model defined by Gisser and Sánchez (1980) to study the effects of strategic behavior on the efficiency of private groundwater exploitation. We have followed Negri's (1989) approach and have evaluated the impact of the strategic externality as the difference between the open-loop and feedback solutions. To compute the feedback equilibria we have used the procedure proposed by Tsutsui and Mino (1990).

Our results show that this procedure does not work for the water pumping differential game developed here. The problem is generated by the *local* nature of the nonlinear strategies. The result is that only the *global* linear strategy leads to a steady-state for the Markov feedback equilibrium of the game. Moreover, we find that strategic behavior, increases the inefficiency of private exploitation with respect to the open-loop equilibrium which captures only the pumping cost externality. However, these results also show that the difference between the socially optimal exploitation and the private exploitation of the aquifer, represented by a feedback equilibrium in linear strategies, decreases with the storage capacity of the aquifer, and thus if this is relatively large the two equilibria are identical for practical purposes as established by Gisser and Sánchez (1980).

Finally, we would like to make some remarks about the scope of this last conclusion. First, as Worthington et al. (1985) and Brill and Burness (1994) have pointed out in two empirical works, the difference between the two regimens may not be trivial if the relationship between average extraction cost and the water table level is not linear and there are significant differences in land productivity or the demand is nonstationary. Consequently, we think that further research is necessary in at least two areas before deciding against regulation of the resource. One should undertake more empirical work to test the hypothesis of linearity, and one could develop more theoretical work to resolve an asymmetric groundwater pumping differential game where the differences in land productivity are taken into account.[20] Moreover, we also believe that using only the firms' profits to characterize socially optimal exploitation is problematic when there is a possibility of irreversible events or irreparable damage to nature. In this case, the water management authority would have to incorporate the water table level into its objective function and postulate some kind of intervention to avoid 'extinction' or the occurrence of irreversible events.[21] This could be another subject for future research. Another situation that could require some kind of regulation arises when groundwater is also used for urban consumption. In this case the water pollution caused by the use of chemical products in agricultural activities alters the quality of water and affects negatively the welfare of urban consumers, generating another externality that could increase the inefficiency of private exploitation of groundwater.

Appendix A

The derivation of H_L is immediate. To calculate H_H we have to solve

$$f'(H) = -\frac{kc_1(N-1)}{N(2N-1)}, \tag{31}$$

which obtained from (18) for $dw/dH = 0$.

Substituting (18) into the Hamilton–Jacobi–Bellman equation yields

$$\begin{aligned} rV(H) &= \frac{1}{2kN(2N-1)}\left[-k(N-1)\left(\frac{g}{k}+c_0+c_1H\right)+\frac{NR}{\gamma-1}\right]^2 \\ &+ \frac{R}{\gamma-1}\left(\frac{g}{k}+c_0+c_1H\right)-\frac{(2N-1)N}{2k}f(H)^2. \end{aligned}$$

Differentiating with respect to H and substituting V' again using Equation (18) for $w(H)$ yields

$$\begin{aligned} & r\left\{\left[\frac{\text{AS}\,N}{(\gamma-1)(2N-1)}\right]\left(\frac{g}{k}+c_0+c_1H\right)+\frac{\text{AS}\,NR}{(2N-1)k(\gamma-1)^2}\right. \\ & \left.-\frac{\text{AS}\,N}{k(\gamma-1)}f(H)\right\} \\ & = \frac{(N-1)c_1}{(2N-1)N}\left[k(N-1)\left(\frac{g}{k}+c_0+c_1H\right)-\frac{NR}{\gamma-1}\right] \\ & + \frac{Rc_1}{\gamma-1}-\frac{(2N-1)N}{k}f(H)f'(H), \end{aligned}$$

which upon rewriting results in

$$f'(H) = \frac{k}{(2N-1)Nf(H)}\left[\frac{\text{AS}\,Nr}{k(\gamma-1)}f(H)-FH-G\right]. \tag{32}$$

On the other hand, we know that at the steady-state $w^* = -R/N(\gamma-1)$, then using Equation (18) one more time, we have

$$f(H^*) = -\frac{k(N-1)}{N(2N-1)}\left(\frac{g}{k}+c_0+c_1H^*+\frac{R}{(\gamma-1)k}\right), \tag{33}$$

and we can use (31), (32) and (33) to obtain H_H.[22]

Appendix B

To calculate H_b we use the linear strategy (21) and the steady-state condition $w^* = -R/(\gamma-1)N$. After substituting F and G, H_b is given by the following expression:

$$H_b = -\frac{R}{kc_1(\gamma - 1)} D_b - \frac{1}{c_1}\left(\frac{g}{k} + c_0\right),$$

where

$$D_b = \frac{\frac{(N-1)kc_1}{N(2N-1)} + \frac{N^2[kc_1(\gamma-1)-rAS]}{(N-1)^2kc_1(\gamma-1)-N^2rAS} y_b}{\frac{(N-1)kc_1}{N(2N-1)} + y_b}.$$

Then the difference between H_b and H^*_{SO} is given by

$$H_b - H^*_{SO} = R\frac{(1 - D_b)rAS - kc_1(\gamma - 1)}{kc_1(\gamma - 1)rAS}, \tag{34}$$

where the denominator is negative.

Let us suppose that

$$(1 - D_b)rAS - kc_1(\gamma - 1) \leq 0, \tag{35}$$

where

$$1 - D_b = \frac{\left(1 - \frac{N^2[kc_1(\gamma-1)-rAS]}{(N-1)^2kc_1(\gamma-1)-N^2rAS}\right) y_b}{\frac{(N-1)kc_1}{N(2N-1)} + y_b}.$$

Operating over (35) we obtain that

$$y_b \geq \frac{kc_1\left[(N-1)^2kc_1(\gamma - 1) - N^2rAS\right]}{N(2N-1)(N-1)\left[rAS - kc_1(\gamma - 1)\right]}, \tag{36}$$

where

$$y_b = \frac{1}{2}\left\{\frac{rAS}{(2N-1)(\gamma-1)} - \left[\left(\frac{rAS}{(2N-1)(\gamma-1)}\right)^2 - \frac{4(N^2kc_1rAS - (N-1)^2k^2c_1^2(\gamma-1))}{N^2(2N-1)^2(\gamma-1)}\right]^{1/2}\right\},$$

after eliminating F by substitution. Then inequality (36) can be rewritten as

$$\frac{rAS}{(2N-1)(\gamma-1)} - \frac{2kc_1\left[(N-1)^2kc_1(\gamma-1) - N^2rAS\right]}{N(2N-1)(N-1)\left[rAS - kc_1(\gamma-1)\right]}$$
$$\geq \left[\left(\frac{rAS}{(2N-1)(\gamma-1)}\right)^2 - \frac{4(N^2kc_1rAS - (N-1)^2k^2c_1^2(\gamma-1))}{N^2(2N-1)^2(\gamma-1)}\right]^{1/2}.$$

Squaring and simplifying terms we get the following contradiction:

$$kc_1(\gamma - 1)r\text{AS}\left[N(N-1) + 2(N-1)^2 - N^2\right] \tag{37}$$

$$-\ (r\text{AS})^2\left[N(N-1) + (N-1)^2\right] \geq 0. \tag{38}$$

Notice that $kc_1(\gamma - 1)r\text{AS} < 0$ and $N(N-1) + 2(N-1)^2 - N^2$ is zero for $N = 2$ and positive for $N > 2$. Given contradiction (38) we have to conclude that the numerator of (34) is positive so that $H_b - H^*_{\text{SO}} < 0$, which establishes that $H_b < H^*_{\text{SO}}$.

Notes

[1] In the early eighties, Gisser published two further papers comparing the optimal and private exploitation of groundwater (Gisser, 1983; Allen and Gisser, 1984). The latter demonstrated that the Gisser–Sánchez rule also applies for the case of an isoelastic demand function.

[2] The cost externality appears because the pumping cost increases with pumping lift, so that withdrawal by one farmer lowers the water table increasing accordingly the pumping costs for all farmers operating over the aquifer. The strategic externality arises from competition among farmers to capture the groundwater reserves through pumping since property rights are not well defined and access is nonexclusive.

[3] This result suggests that if the resource is large the competition among the firms for appropriation of a finite common property resource is feeble, and the strategic externality practically disappears.

[4] If farmers are not price-taking in output market, the demand for irrigation water is given by the marginal revenue product of water which is lower than the value of the marginal product for the same rate of extraction. In this case, we can distinguish three sources of dynamic inefficiency: the cost and strategic externalities and the market power of farmers. This inefficiency has a conservationist bias and could compensate the overexploitation associated with the externalities. Nevertheless, this issue will not be discussed here.

[5] See Knapp and Olson (1995) for a groundwater management model with stochastic surface flows and artificial recharge.

[6] To simplify the notation, the t argument of the variables has been suppressed. It will be used only when necessary for unambiguous notation.

[7] For a formal definition of strategy space and equilibrium concepts used in this chapter, see Tsutsui and Mino (1990). By extension they can easily be adapted to our game.

[8] The transversality conditions are: $\lim_{t\to\infty} e^{-rt}\lambda_i(t) \geq 0$, $\lim_{t\to\infty} e^{-rt}\lambda_i(t)H(t) = 0$, $i = 1, \ldots, N$.

[9] In fact, it is easy to demonstrate that the open-loop Nash equilibrium is symmetric. The proof is available upon request.

[10] In this case we are assuming that a sole ownership (manager or social planner) decides the extraction program for the resource. Under this regime the external effects are *internalized*, all the costs are considered and, consequently, the intertemporal allocation of the resource will be socially optimal.

[11] Finally, we want to comment briefly on how these results affect Gisser and Sánchez's conclusions. Firstly, our results confirm their rule. Secondly, Gisser and Sánchez's estimations are *overvalued* because these authors assume that farmers are *myopic*. If one assumes, as we do, that farmers are *rational*, their private evaluation of user cost will be positive and price will consequently

exceed marginal extraction costs (see (8)). With rational farmers the myopic solution applies only asymptotically, that is, when the number of farmers approaches infinity, $\lim_{N\to\infty} H^*_{\text{OL}} = H^*_{\text{MY}}$.

[12] Remember that stationary Markov strategies describe decision rules that prescribe an extraction rate exclusively as a function of the observed water table, and are, consequently, independent of time.

[13] The derivation of this equation follows step by step the one presented by Dockner and Long (1993) and will not be repeated here.

[14] See Appendix A for the derivation of these two values.

[15] Notice that if we assume that the initial value is lower than H_b and, consequently, lower than the natural hydrologic steady-state, then we are implicitly assuming that there was a previous phase of resource exploitation which led to the initial value we are now assuming. This would mean that our model could only explain the half-history of resource exploitation.

[16] It is easy to show mathematically that this difference is positive. The proof is available upon request.

[17] It is also interesting to point out that a higher return flow coefficient can lighten the existing strategic interdependence among farmers, and, consequently, increase the efficiency of the private exploitation of groundwater. The intuition behind this comment is that a higher coefficient reduces the possibilities of resource appropriation and hence the competition among farmers for capturing the groundwater reserves through pumping.

[18] This happens because the upper bound for the dynamic inefficiency is defined by the myopic solution.

[19] Using data from Gisser and Sánchez (1980), and from Nieswiadomy (1985) we have calculated the upper bound for the dynamic inefficiency defined by (30), and we have found that, for the two cases, the differences in the water elevation are low, both in relative and absolute terms. For the Pecos Basin, New Mexico, the upper bound is 13 ft if the discount factor is 0.1. A discount factor of 0.05 gives approximately 26 ft. For the Texas High Plains, the effect of dynamic inefficiency is less than 5.5 ft for a discount factor of 0.1 and 11 ft if the discount factor is 0.05.

[20] If we assume differences in land productivity we have to rewrite the model considering that the value of water marginal productivity, which defines the individual water demand, is not the same for all farms, and then solve an *asymmetric* differential game. Nevertheless, from the results of our chapter a first conclusion can be drawn: if the Gisser–Sánchez rule applies, the differences in land productivity do not have to affect substantially the efficiency of private exploitation as long as the Gisser–Sánchez rule does not depend on the parameters of the demand function, see expression (30).

[21] See Tsur and Zemel (1995) for the study of the optimal exploitation of groundwater when extraction affects the probability of an irreversible event occurring.

[22] As $H_L < H_H$ the SSL must be below the intersection point of linear strategies in Figure 1. If SSL were above (H', w') then $H_L > H_H$ since H_L is defined by the intersection point of the $\mathrm{d}w/\mathrm{d}H = -\infty$ line with SSL, and H_H is defined by the intersection point of the $\mathrm{d}w/\mathrm{d}H = 0$ line with SSL.

References

Allen, R.C. and Gisser, M. (1984) Competition versus optimal control in groundwater pumping when demand is nonlinear, *Water Resources Research* **20**, 752–756.

Brill, T.C. and Burness, H.S. (1994) Planning versus competitive rates of groundwater pumping, *Water Resources Research* **30**, 1873–1880.

Dockner, E.J. and Long, N.V. (1993) International pollution control: Cooperative versus noncooperative strategies, *Journal of Environmental Economics and Management* **24**, 13–29.

Feinerman, E. and Knapp, K.C. (1983) Benefits from groundwater management: Magnitude, sensitivity, and distribution, *American Journal of Agricultural Economics* **65**, 703–710.

Gisser, M. (1983) Groundwater: Focusing on the real issue, *Journal of Political Economy* **91**, 1001–1027.

Gisser, M. and Sánchez, D.A. (1980) Competition versus optimal control in groundwater pumping, *Water Resources Research* **16**, 638–642.

Kim, C.S., Moore, M.R., Hanchar, J.J. and Nieswiadomy, M. (1989) A dynamic model of adaptation to resource depletion: Theory and an application to groundwater mining, *Journal of Environmental Economics and Management* **17**, 66–82.

Knapp, K.C. and Olson, L.J. (1995) The economics of conjunctive groundwater management with stochastic surface supplies, *Journal of Environmental Economics and Management* **28**, 340–356.

Negri, D.H. (1989) The common property aquifer as a differential game, *Water Resources Research* **25**, 9–15.

Nieswiadomy, M. (1985) The demand for irrigation in the High Plains of Texas, 1957–80, *American Journal of Agricultural Economics* **67**, 619–626.

Provencher, B. and Burt, O. (1993) The externalities associated with the common property exploitation of groundwater, *Journal of Environmental Economics and Management* **24**, 139–158.

Tsur, Y. and Zemel, A. (1995) Uncertainty and irreversibility in groundwater resource management, *Journal of Environmental Economics and Management* **29**, 149–161.

Tsutsui, S. and Mino, K. (1990) Nonlinear strategies in dynamic duopolistic competition with sticky prices, *Journal of Economic Theory* **52**, 136–161.

Wirl, F. (1994) Pigouvian taxation of energy for flow and stock externalities and strategic, noncompetitive energy pricing, *Journal of Environmental Economics and Management* **26**, 1–18.

Wirl, F. and Dockner, E. (1995) Leviathan governments and carbon taxes: Costs and potential benefits, *European Economic Review* **39**, 1215–1236.

Worthington, V.E., Burt, O.R. and Brustkern, R. (1985) Optimal management of a confined groundwater system, *Journal of Environmental Economics and Management* **12**, 229–245.

Demand-Side Technology Standards under Inefficient Pricing Regimes: Are They Effective Water Conservation Tools in the Long-Run?

CHRISTOPHER TIMMINS

1. Introduction

When price-setting policy-makers have objectives other than maximizing social surplus, the conservation potential of demand-side technology standards can be significantly reduced. This is particularly relevant for the analysis of water policy under conditions of scarcity.[1] Regardless of the hydrological or institutional setting, policy-makers around the world have repeatedly demonstrated a preference for non-price demand-side technology standards (like the mandated installation of low-flow toilets or shower heads) or restrictions on use (like rationing or prohibitions on lawn-watering), over efficient pricing in the face of growing water scarcity. Often, these sorts of policies are touted as more equitable than an increase in the price of the most basic of all necessities. This chapter does not dispute the possible equity-enhancing benefits of non-price policies for water conservation, compared to efficient pricing. Rather, it focuses on the overall effectiveness of these policies in dealing with problems of water scarcity, particularly when water management involves dynamic considerations.

The most water-scarce regions of the world often rely on groundwater stocks to serve the basic needs of residential, commercial, and agricultural users. In these settings, the water-pricing decisions of political authorities have intertemporal consequences: underpricing and, consequently, overextracting water today will reduce available water stocks tomorrow, increasing future costs of extraction. Still, when confronted with these dynamic costs, politicians and water managers have repeatedly found it optimal to price water below even its static marginal cost of provision. Consider the example of California's arid Central Valley. Table 1 describes the marginal price and price mark-down (averaged over time and calculated without accounting for dynamic or external cost implications of current extraction) for 13 groundwater dependent cities in that region. Every city exhibits a statistically and economically significant discount of price below the cost of provision at the margin. This imposes an allocational loss of efficiency on the residents of these

P. Pashardes et al. (eds.),
Current Issues in the Economics of Water Resource Management, 123–146.

communities. Resources expended in the provision of water would yield a greater net benefit in other uses.

Decisions to underprice water and incur these deadweight losses typically arise out of equity or political concerns, but are influenced by the prevailing conditions of water scarcity. While the short-run implication of a mandatory low-flow appliance installation policy may be a reduction in consumption and mitigation of water scarcity, the long-run implications might involve a price reaction to the reduced scarcity that offsets much of the policy's original conservation potential. The application of such a policy, in essence, frees the water authority to pursue even more aggressive underpricing policies such that, in the long-run, very little water is actually saved.

This chapter illustrates the potential for this problem using data on the 13 California cities listed in Table 1 between 1970 and 1992. First, a model is specified that explains the decisions of the managers of these cities to underprice their water relative to the static marginal cost of provision, even when confronted by the additional dynamic costs associated with aquifer stock depletion. Second, the parameters of the municipal manager's objective function, which describe this price discounting behavior, are recovered econometrically with a dynamic fixed-point algorithm nested inside a simulated maximum likelihood estimation procedure. Finally, the estimated dynamic objective function of the municipal manager is used in counterfactual experiments to uncover the short- and long-run conservation potential of a policy mandating the installation of a demand-side technological standard, such as ultra-low-flow toilets or low-flow shower heads. The implications of such a policy are compared under alternative pricing regimes, and price is demonstrated to be a cost-effective policy for achieving a particular level of groundwater conservation. Assuming that the non-price policy is preferable on grounds of equity, however, the simulation demonstrates that the long-run price reaction of municipal regulators wipes-out 80% of the potential conservation gain of the policy. The implication for policy-makers is that non-price demand management policies will not be effective unless accompanied by at least a modicum of pricing austerity.

The chapter proceeds as follows. Section 2 describes a number of water-scarce parts of the US where political authorities have opted for technological standards over price-based demand management policies, and reviews the capabilities of low-flow technologies to reduce water consumption (both with and without subverting behavior on the part of homeowners). Section 3 describes a flexible model of water manager decision-making that explains observed, inefficient price-discounting decisions as the result of optimizing behavior. Section 4 briefly outlines an estimation technique that econometrically identifies the parameters of the objective function described in Section 3, discusses the panel of California municipal data used to recover those parameters, and specifies functional forms to be used in estimation and simulation. An Appendix provides more detail on the estimation algorithm, while a complete discussion of the model, data, and estimation algorithm can be found in Timmins (2002). Section 5 uses counterfactual simulations to demonstrate the long-run conservation implications of a technology standard given the pricing behavior of municipal managers estimated in Section 4, and Section 6 concludes by suggesting implications for policy-makers.

2. Low-Flow Toilets: The Preferred Policy Tool of Municipal Water Managers

Approximately two-thirds of urban water use in the US is attributable to residential consumers, and of the water that they use indoors, most is employed by a small number of household appliances (i.e., 36% in the toilet, 28% in the bath or shower, and 20% in the laundry) (California DWR, 1993, p. 153). It is therefore not surprising that politicians, averse to using price as a policy tool in the face of growing water scarcity, have instead promoted the adoption of technological standards aimed at reducing these sources of water use in the home. For example, between 1989 and 1992 (years in which much of the western US was experiencing severe droughts), 17 states either passed or considered laws to mandate the use of ultra-low-flow toilets, while similar laws have been passed at the municipal level in Maui, San Francisco, San Diego, Los Angeles, Denver and Washington, DC (Frammolino, 1992). At the national level, the US Congress passed the Energy Policy Act of 1992, part of which required all new and remodeled commercial buildings and residential homes to have low-flow toilets and shower heads (*The Columbus Dispatch*, 1999). Often, these policies do not take the form of technological mandates, but rather subsidies paid to homeowners for the voluntary adoption of a low-flow device. In San Diego, for example, the city council approved 35,000 rebates of approximately $85 each to any homeowner who would install an ultra-low-flow toilet, while the city of La Verne offered $50 to anyone who would do the same and $100 to anyone who would adopt conservation landscaping practices (*The Los Angeles Times*, 1991). By 1996, similar rebate programs had been responsible for financing the installation of over 136,000 low-flow toilets nationwide (La Rue, 1995).

Policy-makers confronted with water scarcity have repeatedly revealed a preference for conservation retrofit plumbing kits and low-flow toilets and fixtures over price increases. Often, their professed reasons for this preference involve equity considerations; i.e., concerns about raising a regressive charge on a necessity (Lilly, 1992). In other contexts, such equity concerns appear to be clouded by political motivations. Timmins (2000) demonstrates, in a cross-section of California cities, that the lower the income of the median voter, the more will relatively progressive property and sales taxes be used to substitute for regressive water charges in municipal revenues.

The unwillingness of politicians to increase prices in the face of rising water scarcity is not limited to the US. In Cyprus, for example, residential users currently pay a price equal to approximately 66% of the cost of water provision, while the country's total demand for water has exceeded its supply in three out of the last five years (Socratous, 2000). Still, the Parliament has refused repeatedly over the last six years to pass a proposal to increase the price of urban water. While raising price continues to be an unpopular alternative for water conservation in this arid country, there are currently calls for increased subsidies for the adoption of new water-conserving technologies (Socratous, 2000).

With their popularity among policy-makers well-established, it is important to document whether low-flow appliances are capable of reducing residential water use. From a technical perspective, the answer to this question is straightforward. Estimates suggest that a typical home can cut its average daily water use from 74 to 52 gallons with the installation of these devices, and that the appliances installed in just new US homes in 1998 could save 16 billion gallons of water nationwide (Johnson, 1999). In southern California, where adoption of low-flow toilets and shower heads has been undertaken in new and old homes alike, the devices are purported to be capable of saving the region more than 14 billion gallons of water a year (Bernstein, 1996). Factoring in the behavior of the consumer, however, the answer to this question is not clear. 'Double-flushing' has often been cited as a potential failure of low-flow toilets (Frammolino, 1992), and anecdotal evidence suggests that homeowners have found numerous ways to alter mandatorily-installed commodes to use more water (Booth, 1999). Often, these alterations are prompted by problems arising from clogged sewers, which were originally constructed to operate with greater water flow (Frammolino, 1992). These problems typically lead to additional costs in terms of plumbing fees and property damage.[2] With these potential problems in mind, low-flow technologies have improved rapidly over the last ten years (as have new styles in conservation landscaping), making their technical potential for water conservation very real.

With all of this technical potential, however, low-flow toilets and showerheads have proven most effective when adopted in conjunction with increasing prices for water at the margin; i.e., when they are chosen willingly by consumers as an optimal response to prices that reflect water scarcity. Only with such prices would we expect consumers to avoid illegal alterations designed to make the low-flow devices function like traditional fixtures. This has clearly been the case in Los Angeles, where the adoption of increasing block rate price schedules (e.g., a household's marginal price triples if usage exceeds 13,000 gallons a month) in conjunction with the subsidized adoption of low-flow toilets and showerheads resulted in a 12% reduction in water use between 1989 and 1995, despite a growing urban population (Bernstein, 1996). Similar benefits resulted from price-conscious consumers adopting conservation landscaping practices and plastic swimming pool covers to prevent evaporation[3] (Bernstein, 1996). On the other hand, consumers

Table 1. Marginal cost and marginal price: averaged over time for 13 California cities.

City	Obs	Average marginal cost	Average marginal price
Clovis	16	19.67***	100.78
Delano	15	118.60*	33.46
Dinuba	15	97.69*	170.12
Exeter	22	47.74*	61.77
Firebaugh	14	174.40*	0.00
Fresno	19	124.99*	0.00
Hanford	21	65.15*	69.02
Kerman	6	109.44**	0.00
Madera	16	124.46*	0.00
Mendota	15	101.50*	104.44
Reedley	17	171.14*	0.00
Sanger	16	43.88*	60.19
Shafter	15	120.61*	0.00

* indicates significance at the 0.5% level; ** indicates significance at the 2.5% level, and *** indicates statistical significance at the 10% level. All figures are reported in constant 1982–1984 dollars.

not faced with the proper price incentives have consistently found ways to subvert inconvenient mandated technologies, with the result being lost conservation potential.

The remainder of this paper examines a corresponding problem which occurs on the other side of the market for water – the potential for inefficient *supply-side* price responses to the institution of demand-side technology standards. Such pricing behavior, as will be shown, can wipe-out most of the conservation potential of these technologies in the long-run.

3. The Model

3.1. A Flexible Static Objective

In endeavoring to explain the pricing decisions of water managers like those in charge of the California cities listed in Table 1, we clearly cannot appeal to a profit-maximizing or even a net-surplus-maximizing objective function. According to a conservative, static definition of the cost of providing water (i.e., a definition that ignores the future costs of aquifer depletion and other current and future external costs of over-extraction), these managers have consistently chosen to price water at a discount, inducing a loss of social surplus. One possible explanation, which coincides with the political reality in these cities (i.e., where water managers are the appointees of elected city officials), is that managers care directly about the surplus of their water consuming constituents:

$$CS(P, \mathbf{X}, \varepsilon^d; \boldsymbol{\theta}_1) = \int_P^\infty D(s, \mathbf{X}, \varepsilon^d; \boldsymbol{\theta}_1)\,\mathrm{d}s, \tag{3.1}$$

where $D(P, \mathbf{X}, \varepsilon^d; \boldsymbol{\theta}_1)$ describes aggregate municipal water demand as a function of (i) marginal price (P), (ii) exogenous demand determinants ($\mathbf{X}$), such as the number of households (HOUSE), expected rainfall (μ_R), and the median level of real income in the city (INC),[4] (iii) determinants of demand that are unforseen at the time when the manager sets prices at the start of the fiscal year (ε^d), such as water used for fighting fires, and (iv) a vector of parameters ($\boldsymbol{\theta}_1$).[5] If the municipal management cared only about the surplus of its water-consuming constituents, it would set the marginal price to zero; this is, in fact, often the case in California's Central Valley, where many cities do not have water meters (see Table 1). Many cities do meter water use, however, and charge for consumption at the margin. This implies an additional explanation. Peltzman (1971) suggests that municipal providers of commodities like water may view their taxpaying constituents as 'shareholders' in the municipality, meaning that profits earned from the provision of water would raise the value of the city's 'stock'. This, in turn, raises the likelihood that shareholders will vote to keep the management (e.g., the mayor or town council) in office. In addition to consumer surplus, municipal managers might therefore care directly about profits from operation:

$$\begin{aligned}\pi(P, \mathbf{X}, h, \varepsilon^d, \varepsilon^c; \boldsymbol{\theta}_1, \boldsymbol{\theta}_2)\\ = P \times D(P, \mathbf{X}, \varepsilon^d; \boldsymbol{\theta}_1) - C(D(P, \mathbf{X}, \varepsilon^d; \boldsymbol{\theta}_1), h, \varepsilon^c, \boldsymbol{\theta}_2),\end{aligned} \tag{3.2}$$

where the first term refers to variable revenues and the second term refers to the cost of water provision; i.e., the cost of groundwater extraction (which constitutes approximately one-third of the total cost of groundwater provision), in addition to the costs of treatment and distribution. The cost of groundwater pumping is expected to be an increasing function of the quantity extracted and a decreasing function of the available stock, quantified by the inverse measure, lift-height (h).[6] ε^c represents determinants of the variable costs of water provision that are unknown to the water manager at the time when prices are being set, and $\boldsymbol{\theta}_2$ represents a vector of cost parameters. Fixed costs of water provision are poorly described in available data and are assumed to be recouped with fixed service charges, which are also absent from Equation (3.2).

Municipal managers are assumed to maximize some combination of the benefits to their constituents in their dual roles as water consumers and taxpayers:

$$\begin{aligned}\Pi(P, \mathbf{X}, h, \varepsilon^d, \varepsilon^c; \nu, \boldsymbol{\theta}_1, \boldsymbol{\theta}_2)\\ = \nu \times CS(P, \mathbf{X}, \varepsilon^d; \boldsymbol{\theta}_1) + (1 - \nu) \times \pi(P, \mathbf{X}, h, \varepsilon^d, \varepsilon^c; \boldsymbol{\theta}_1, \boldsymbol{\theta}_2),\end{aligned} \tag{3.3}$$

where the parameter ν determines the weight placed on the welfare of each group.

3.2. DYNAMICS

Municipal managers in the area of California being considered rely on renewable aquifer stocks for the raw inputs used to provide water to municipal consumers. Their decision process is therefore more complex than simply choosing P at the start of each fiscal year to maximize the payoff described in (3.3) in expectation. Managers instead choose an infinite series of prices to maximize the expected present discounted value of the future stream of payoffs from providing municipal water, subject to a stochastic law of motion describing the lift-height to the groundwater. Furthermore, they do this subject to a non-negativity constraint on price:

$$\max_{[P_t \geq 0]} \quad E_{\Phi_0}\left[\sum_{t=t_0}^{\infty} \beta^{t-t_0} \Pi_t(P_t, \mathbf{X}_t, h_t, \varepsilon_t^d, \varepsilon_t^c; \nu, \boldsymbol{\theta}_1, \boldsymbol{\theta}_2)\right]$$

$$\text{s.t.} \quad h_{t+1} = f_h(h_t, \mathbf{X}_t, D(P_t, \mathbf{X}_t, \varepsilon_t^d, \varepsilon_t^c; \boldsymbol{\theta}_1), \varepsilon_t^S; \boldsymbol{\gamma}),$$

$$\mathbf{X}_{t+1} = f_X(\mathbf{X}_t, \varepsilon_t^X, \boldsymbol{\zeta}), \tag{3.4}$$

where $\boldsymbol{\gamma}$ represents a parameter vector used to describe the law of motion for h, $\boldsymbol{\zeta}$ represents a parameter vector used to describe the law of motion for $\mathbf{X}$, and Φ_0 represents the set of information available at t_0.[7] $\beta \in [0, 1]$ measures the manager's discount factor for future payoffs.

A manager who places a great deal of weight on the welfare of his constituents in their role as consumers of water would choose a low price path, but would be tempered by the knowledge that very low prices would deplete stocks, raising future extraction costs and necessitating higher future prices, which would hurt water consumers. It is this dynamic that will drive the inefficient long-run price response to the institution of a demand-side technology standard, which proves damaging for water conservation.

The dynamic objective described in Equation (3.4) suggests the following Bellman functional form, which describes the year-to-year iterative problem confronting the municipal manager:

$$V(h_t, \mathbf{X}_t; \boldsymbol{\Theta}) = \max_{[P_t \geq 0]} E_{\varepsilon_t^d, \varepsilon_t^c}\big[\Pi(P_t, \mathbf{X}_t, h_t, \varepsilon_t^d, \varepsilon_t^c; \boldsymbol{\Theta}) + \beta E_{\varepsilon_t^h, \varepsilon_t^X}[V(h_{t+1}, \mathbf{X}_{t+1}; \boldsymbol{\Theta})]\big]. \tag{3.5}$$

Stokey and Lucas (1989) describes the conditions under which Equation (3.5) constitutes a contraction mapping, such that this dynamic problem yields a stationary policy function, $P(h, \mathbf{X}; \boldsymbol{\Theta})$. This policy function describes the optimal pricing behavior pursued by a manager given a stock of available water (h), exogenous municipal attributes ($\mathbf{X}$), and a vector of parameters ($\mathbf{\Theta}$). The full parameter vector includes the manager's preference parameter (ν), the parameters of the demand and extraction equations ($\boldsymbol{\theta}_1$, $\boldsymbol{\theta}_2$), and the parameters of the distributions underlying the stochastic sources of uncertainty (ε^d, ε^c, ε^S, ε^X) over which the manager integrates when determining price. Solving for this policy function is complicated by the fact that the non-negativity constraint on price binds very often in the data (see Table 1), implying that Euler equation techniques cannot be employed. Instead, a numerical algorithm is used to solve for the value function that is the fixed point of the contraction mapping described by (3.5). The corresponding policy function is described by a flexible spline representation that can explicitly represent the non-negativity constraints on price.

3.3. The Importance of a Flexible Objective

Water price discounts impose costs on taxpayers. If water authorities choose to discount price and incur the wrath of taxpayers at the voting booth, it must be because they find discounts to water consumers to be valuable. This could be because of noble equity concerns, or self-interested desires to be re-elected by a majority of low-income constituents. In either case, if we are going to model the long-run price response of a municipal water authority to the imposition of a demand-side technology standard, we need to first recover price-setting managers' preferences for price discounts, which are summarized by their preferences for the welfare of different constituent groups.

4. Data and Estimation

Timmins (2002) provides a detailed discussion of a data panel describing 13 groundwater dependent cities in California's Central Valley between 1970 and 1992, and an estimation algorithm that uses those data to recover estimates of municipal managers' preferences for the welfare of their competing constituent groups. Section 4.2 briefly describes the data used in the calculation of the likelihood function. Section 4.1 first describes the specific functional forms used in the estimation procedure and in the simulations in Section 5. It is important to note that the general results concerning the impact of the long-run price response on the conservation potential of demand-side technology standards do not depend on any of these assumptions. They are made only to deal with the specific strengths and shortcomings of the available data. Section 4.3 describes the results of this estimation procedure that are most relevant for the simulations performed in Section 5.

4.1. Functional Forms and Estimation Intuition

A semilog specification is used to model aggregate demand in city i, year t. This functional form accounts for the fact that demand does not become infinite at a marginal price of zero (an event which is frequently observed in the data), as well as for the fact that there is not a reasonably priced backstop technology to replace urban water.

$$\ln D_{i,t} = \delta_0 + \boldsymbol{citdum}'\delta_{0,i} - \delta_1 P_{i,t} + \delta_2 \mathrm{INC}_{i,t} - \delta_3 R_{i,t} + \delta_4 \mathrm{HOUSE}_{i,t} + \varepsilon^d_{i,t},$$

$$R_{i,t} = \mu_{R,i} + \varepsilon^R_{i,t},$$

$$\begin{bmatrix} \varepsilon^R_{i,t} \\ \varepsilon^d_{i,t} \end{bmatrix} \sim \text{i.i.d. } N\left(\begin{bmatrix} 0 \\ 0 \end{bmatrix}, \begin{bmatrix} \sigma^2_R & 0 \\ 0 & \sigma^2_d \end{bmatrix} \right). \tag{4.1}$$

citdum represents a vector of city indicator dummies, and $R_{i,t}$ represents annual rainfall, which is unknown by municipal managers at the start of the fiscal year when price is chosen, but which is distributed as white-noise around a city-specific mean.

Given the imperfect reporting of variable production costs described in the following subsection, operating profits are specified as a function of marginal price (P), a time-invariant, city-specific component of non-extraction costs per acre-foot[8] of water produced (c), a Cobb–Douglas specification of the costs of groundwater extraction ($C^g(\cdot)$), and an unreported component of net revenues per acre-foot (ξ):

$$\pi(P, \mathbf{X}, h, \xi, \varepsilon^d, \varepsilon^g; \boldsymbol{\theta}_1, \boldsymbol{\theta}_2)$$
$$= (P - c + \xi) \times D(P, \mathbf{X}, \varepsilon^d; \boldsymbol{\theta}_1) - C^g(D(P, \mathbf{X}, \varepsilon^d; \boldsymbol{\theta}_1), h, \varepsilon^g; \boldsymbol{\theta}_2), \tag{4.2}$$

where

$$\ln C^g(D(P, \mathbf{X}, \varepsilon^d; \boldsymbol{\theta}_1), h, \varepsilon^g; \boldsymbol{\theta}_2)$$
$$= \boldsymbol{citdum}'\alpha_{0,i} + \alpha_1 \ln h_{i,t} + \alpha_2 \ln D_{i,t} + \varepsilon^g_{i,t}. \tag{4.3}$$

ξ is intended to measure systematic under- or over-reporting of marginal costs or revenues, and is allowed to be serially correlated over time.

$$\xi_{i,t+1} = \rho_0 + \rho_1 \xi_{i,t} + \varepsilon^\xi_{i,t}, \quad \varepsilon^\xi_{i,t} \sim \text{i.i.d. } N(0, \sigma^2_\xi). \tag{4.4}$$

In fact, ξ is treated in the estimation algorithm as an unobserved state variable, along with the observed state variable, h, which evolves according to:

$$h_{i,t+1} = \boldsymbol{citdum}'\delta_{0,i} + \delta_1 h_{i,t} + \delta_2 D(P_{i,t}, \mathbf{X}_{i,t}, \varepsilon_{i,t}; \boldsymbol{\theta}_1) + \delta_3 \mathbf{X}_{i,t} + \delta_4 Z_{i,t} + \varepsilon^h_{i,t},$$

$$\varepsilon^h_{i,t} \sim \text{i.i.d. } N(0, \sigma^2_h), \tag{4.5}$$

where $Z_{i,t}$ measures the effect on aquifer stock of groundwater extraction by nearby farmers, and is described in detail in Timmins (2002).

We now re-specify the dynamic programming problem of the water manager in city i (subscript i suppressed here and in the following discussion):

$$V(h_t, \xi_t \mid \mathbf{X}_t, \boldsymbol{\Theta}) = \max_{[P_t \geq 0]} E_{\varepsilon_t^d, \varepsilon_t^g}\Big[\Pi(P_t, \mathbf{X}_t, h_t, \xi_t, \varepsilon_t^d, \varepsilon_t^g, \boldsymbol{\Theta}) + \beta E_{\varepsilon_{t+1}^h, \varepsilon_{t+1}^\xi}[V(h_{t+1}, \xi_{t+1} \mid \mathbf{X}_t, \boldsymbol{\Theta})]\Big]. \tag{4.6}$$

$\mathbf{X}_t$, which measures exogenously evolving attributes of each municipality (e.g., population, real income, mean rainfall) is assumed to be treated myopically by municipal managers in order to reduce the size of the state space and make the problem computationally tractable for the purposes of estimation. Given the stationarity or slow progress of these attributes over time, this assumption will not affect the results of the estimation significantly, but given expanded computing resources, it should certainly be relaxed. Under this assumption, $\mathbf{X}_t$ does not appear in the state vector, but is instead a conditioning variable both in the current and next-period value function.[9] As demonstrated in Timmins (2002), this conditional dynamic programming problem yields a time-stationary policy rule (conditional upon $\mathbf{X}_t$), $P(h, \xi \mid \mathbf{X}, \boldsymbol{\Theta})$, given a vector of parameters $\boldsymbol{\Theta}$. Conditional upon observed groundwater stock (h) and municipal attributes ($\mathbf{X}$), this conditional policy rule is a function of the stochastically determined unreported net revenues per unit (ξ), which allows us to determine the probability of observing any particular value of price.

The estimation algorithm employs a maximum likelihood procedure that uses the simulated probabilities of observing combinations of lift-height, aggregate consumption, and price, given the predicted pricing behavior arising from the dynamic objective function described in Equation (4.6). For a particular vector of parameters ($\boldsymbol{\Theta}$), these probabilities are recovered for each observation in the data panel, and are combined in a likelihood function with the probability of observing a series of unreported net marginal revenues (ξ) that would be consistent with the observed pricing behavior. The likelihood function is then maximized over the space of $\boldsymbol{\Theta}$. This estimation procedure is described in more detail in the Appendix.

4.2. Data

Recovering econometric estimates of municipal managers' preferences requires a great deal of data. As described above, it necessitates coordinated production, aquifer, climate, and demographic data on numerous municipalities over time. I have constructed such a data panel, which describes 13 cities in California's southern San Joaquin Valley between 1970 and 1992. The variables in this panel fall into the following categories: (i) *characteristic data*, which describe the exogenous market conditions under which municipal managers operate, (ii) *economic decision data*, which describe the endogenously determined policy options pursued by municipal managers and the resulting market equilibria, and (iii) *environmental data*, which describe the complex hydrogeological and climatic forces that influence municipal managers' decisions.

The characteristic data consist of variables that describe the markets served by municipal utilities but which are assumed (i) to evolve exogenously from municipal managers' decisions, and (ii) to be treated myopically by municipal managers. Among these variables are key components of municipal demand, including the number of households (HOUSE) and the real income of urban residents, adjusted to account for a non-linear price schedule (INC). From the California State Department of Finance, I have acquired annual population estimates for each of the 13 municipalities listed in Table 1 (California Department of Finance, 1970–1993), and from decennial US Census data on household incomes, in combination with annual California Department of Finance statistics on per-capita incomes, I have constructed imputed real annual median household incomes by city.

Economic decision data are composed of those variables that describe the decisions made by municipal water utility managers and urban consumers of water, as well as the market equilibria that those decisions, in conjunction with characteristic and environmental variables, induce. The most important of these data describe the residential rate structures set by municipalities. In particular, data on total revenues, service charges, and block-rate pricing structures, which are obtained from the California State Controller's Office: *Municipal Income and Expense Statements*, are used to construct approximate two-part municipal pricing measures that are observed by the econometrician with error. Aggregate municipal demand is sporadically observed in State Controller records, which are supplemented by California Department of Water Resources databases and detailed pumping records that have been obtained directly from municipal water utility management. Other economic decision variables describe operating expenditures. The State Controller's records decompose the variable costs of water utility operation to a level commensurate with the model outlined above. In particular, pumping costs comprise over 32% of all variable expenditures in water production. Data describing the costs of water treatment, pressurization, transmission and distribution, as well as general administrative and sales & customer account maintenance costs, detail the total expenditures made in each of these activities as raw water is processed and delivered to consumers. Along with the pumping cost and rate structure data, these figures allow for the direct measurement of variable profits (i.e, a key component of the municipal manager's hypothesized objective function), but are also a significant source of measurement error in the model.[10]

Finally, environmental data describe climatological and hydrogeological determinants of behavior. From California's Department of Water Resources, Fresno Bureau, I have obtained detailed hydrogeological panel data that characterize the San Joaquin Valley aquifer, the only source of raw water for the 13 cities in the data panel, at the level of a six-square-mile grid between 1970 and 1993. By integrating this data with municipal altitude data, I am able to determine a city's lift-height each year, which is a convenient measure of its groundwater stock and an important determinant of its variable costs. These data are also used to determine how each municipality might be affected by the extraction decisions of its neighbors ($Z_{i,t}$), allowing any 'common-pool' aspects of the aquifer to be handled explicitly.[11] Monthly rainfall figures for several weather stations located across the Valley were obtained from the National Climatic Data Center; each city's rainfall is approximated by that recorded at the nearest weather station.

4.3. Estimation Results

Applying the estimation algorithm in Section 4.1 to the data described in Section 4.2 yielded the following parameter estimates, presented in the form of the equations described in Section 3.[12] Coefficients describing city-specific heterogeneity have been omitted for the sake of brevity. Standard errors appear in parentheses.

Demand:

$$\ln D_{i,t} = \underset{(0.170)}{7.268} - \underset{(0.003)}{0.749} P_{i,t} + \underset{(9.70 \times 10^{-6})}{4.03 \times 10^{-5}} \text{INC}_{i,t} - \underset{(5.40 \times 10^{-5})}{1.20 \times 10^{-4}} \mu_R + \underset{(1.40 \times 10^{-5})}{1.70 \times 10^{-4}} \text{HOUSE}_{i,t}$$

Extraction Cost:

$$\ln C^g_{i,t} = \underset{(0.890)}{-1.656} + \underset{(0.180)}{1.085} \ln h_{i,t} + \underset{(0.071)}{1.175} \ln D_{i,t}$$

Lift-Height Law of Motion:

$$h_{i,t} = \underset{(0.013)}{0.972} h_{i,t-1} + \underset{(2.550)}{0.801} + \underset{(2.00 \times 10^{-4})}{6.35 \times 10^{-4}} D_{i,t} - \underset{(0.001)}{0.010} R_{i,t-1} - \underset{(0.180)}{0.768} Z_{i,t}$$

Unreported Net Marginal Revenues:

$$\xi_{i,t+1} = \underset{(24.330)}{-19.999} + \underset{(0.230)}{0.580} \xi_{i,t} \qquad \sigma_\xi = \underset{(28.460)}{134.995}$$

Statistically significant estimates of the parameters of the distributions of the stochastic elements of the model, $\varepsilon^d_{i,t}$, $\varepsilon^R_{i,t}$, $\varepsilon^g_{i,t}$, $\varepsilon^h_{i,t}$, and $R_{i,t}$, were calculated as well but are not reported. Average costs per acre-foot other than those incurred extracting groundwater (c) were also generally statistically significant and varied across cities from 110.826 to −32.24, indicating that there is a wide disparity in the cost structures of the firms in the sample. The municipal manager's preference parameter, ν, is estimated to be 0.73 with a standard error of 0.005, rejecting the hypotheses that the manager sets price to maximize dynamic profits ($H_0 : \nu = 0$) or social surplus ($H_0 : \nu = 1/2$).

5. The Implications of Pricing Responses to Demand-Side Technology Standards

In this section, the point estimates recovered in the previous section are used to calibrate a simulation model of a mandatory low-flow appliance adoption policy. A simulation model will prove more useful than simply examining actual low-flow appliance adoption policies, because it will allow us to consider the outcomes of alternative pricing responses on the part of municipal managers. First, the conservation and welfare consequences of such a policy are analyzed while allowing municipal managers to adjust prices in the manner they see fit. These results are then compared to the conservation and welfare outcomes of a similar demand-side standard policy instituted under the constraint that managers price as they would have in the absence of a mandatory low-flow appliance adoption policy. The difference between the outcomes of these simulations describes the lost conservation potential of the appliances owing to inefficient price responses of managers. Next, demand-side standards are shown to be cost-ineffective policy tools for groundwater conservation; i.e., under a reasonable set of assumptions, the same conservation potential could be achieved by a small consumption tax (applied to the status quo pricing schedule) with a preferable outcome for social surplus.

5.1. Simulation Methodology

In conducting a simulation of aggregate demand and lift-height under a continuation of the current pricing behavior, $\mathbf{X}_{i,t}$ is first forecast 50 years into the future via time series extrapolation. Given the structural parameter estimates found above, an assumed value for β (i.e., 0.95), and a set of initial conditions used to describe a representative municipality, the pricing decisions of that city's manager and the subsequent lift-height realizations are then simulated over the 50-year horizon. Recognizing that $\mathbf{X}_{i,t}$ and all stochastic processes enter the time-paths of price and lift-height non-linearly, these dynamic simulations are repeated for 180 sets of independent random draws on $\mathbf{X}_{i,t}$. Without including a mandatory low-flow appliance adoption policy, the resulting sets of simulated paths are averaged to produce the mean predicted 'status quo' price and lift-height series for the representative city.

In order to simulate the impact of mandatory low-flow appliance adoption, beginning in the first simulation year, each household is assumed to need 15% less water in order to enjoy the same consumer surplus from water use as prior to the policy.[13] Given the problems with low-flow appliances cited by homeowners in

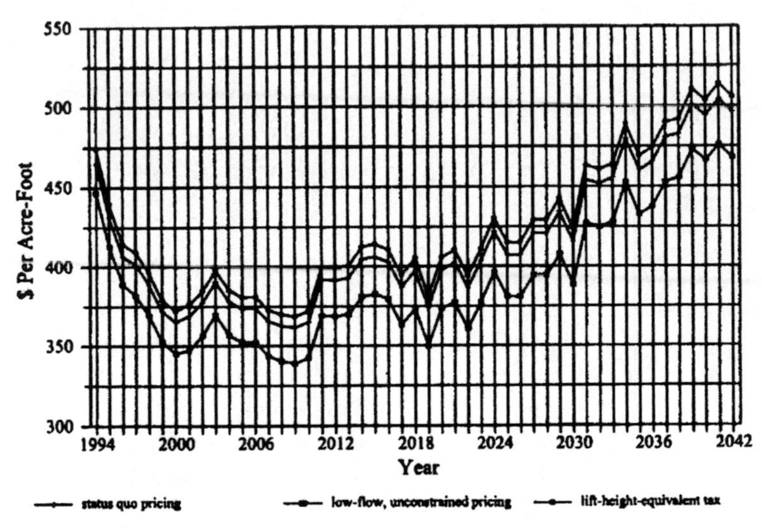

Figure 1. Price per acre-foot: status quo *versus* low-flow installation.

Section 2, this obviously paints a rosy picture of a mandatory low-flow appliance adoption policy. The low-flow technology is assumed to cost $50 per household, which is also an understatement. Costs are assumed to be borne by the government, which is equivalent to their being paid by taxpayers given the model described above, and they are assumed to decline each year at a rate of 2.5%. In the first simulation year, all existing households are retrofitted, and in later years, all new houses adopt the low-flow technology. With these additional features added to the model, the same simulation procedure used above to derive the status quo price schedule is used to calculate an 'unconstrained mandatory low-flow adoption' price schedule.

A final set of simulations is used to describe the behavior of the aquifer stock over time given that the mandatory low-flow appliance installation policy is implemented, but municipal managers are constrained to price according to the status quo price schedule. The same simulation technique described above is used, only the status quo price schedule from the first set of simulations is read into the simulation model as data.

5.2. Low-Flow Appliances and Pricing Behavior: Conservation Potential

Figure 1 illustrates the mean simulated price paths under (i) status quo pricing, (ii) unconstrained mandatory low-flow adoption pricing, and (iii) a modest tax on water consumption, which is described below. On average, over the 50-year simulation horizon, the municipal manager chooses to markdown price by $24, or 5.8% of the rate under status quo pricing, in response to the mandatory implementation of

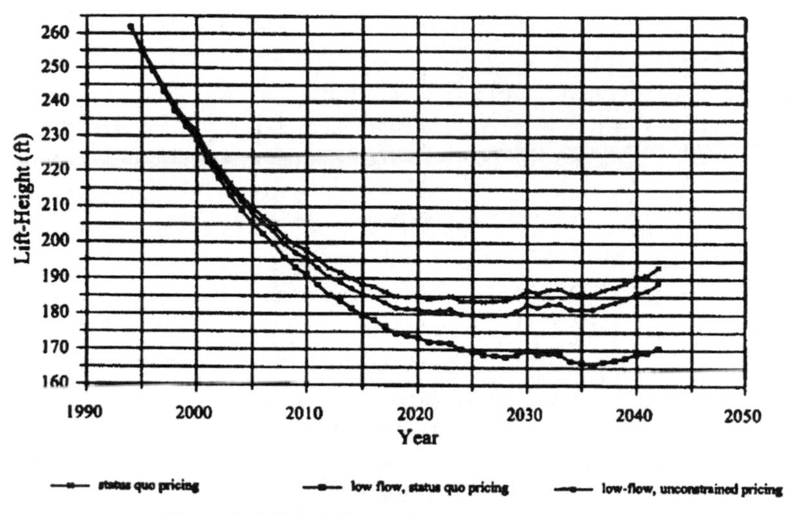

Figure 2. Lift-height by alternative pricing policy.

low-flow appliances. Moreover, this is a conservative prediction of the markdown owing to the assumption that consumer surplus does not fall at all in response to the conversion to low-flow appliances (holding prices constant). This sizable markdown has important consequences for the stock of groundwater.

Figure 2 describes the behavior of lift-height under (i) status quo pricing, (ii) unconstrained mandatory low-flow adoption pricing, and (iii) constrained mandatory low-flow adoption pricing policies. Without any constraints on pricing behavior, the mandatory introduction of low-flow devices yields a 50-year lift-height of 188.82 feet, 4.52 feet less than the 50-year lift-height under status quo pricing. This result is in sharp contrast to the 50-year lift-height under mandatory low-flow appliance adoption with prices constrained to the status quo schedule; i.e., 170.80 feet. Approximately 80% of the long-run conservation potential of these devices is lost to the price response of municipal managers.

5.3. Low-Flow Appliances Versus a Water Tax: Cost-Effectiveness

Having established that a large percentage of the potential conservation gains from mandating the introduction of low-flow appliances will be eliminated by inefficient price responses on the part of municipal managers, we can next ask whether such a policy is even cost-effective. Might the same conservation gains be achieved with a tax on water use, and yield a relative improvement in social surplus (i.e., the surplus enjoyed by water consumers in combination with that of taxpayers) in the process? Considering average surplus changes per household per year, consumer surplus under unconstrained mandatory low-flow appliance adoption pricing either rises by $13.98 (if we assume that consumer surplus does not fall at all with the water-use reductions implied by the low-flow devices) or falls by $3.81 (if consumer surplus is assumed to fall along with reductions in water use). In either case, average operating profits per household per year fall by $7.92. This implies a net increase in social surplus of $6.06 per household per year if low-flow appliances are not detrimental to consumer surplus and a decline of $11.72 if they are detrimental.

To contrast these results with a lift-height-equivalent price policy, a time-invariant water-use tax, τ, was considered. Managers were constrained to set prices, P_t, according to the status quo price schedule, and alternative values of the tax were considered until the after-tax price schedule, $P_t' = P_t(1+\tau)$, yielded a 50-year lift-height equivalent to that achieved by mandatory low-flow appliance adoption with unconstrained pricing (i.e., 188.82 feet). A tax rate of $\tau = 1.95\%$ was required. Under this tax, consumer surplus per household was found to fall by $2.98 each year relative to status quo pricing alone, while operating profits per household rose by $7.54 every year, for a net annual increase in social surplus per household of $4.56. It is only under the most optimistic of assumptions regarding the effects of mandatory appliance adoption on consumer surplus that the water tax is not a relatively cost-effective policy for achieving a 4.52 foot reduction in 50-year-ahead lift-height.

6. Conclusions

The simulations described in Section 5 are clearly stylized exercises. We would not expect, for example, to see any municipal manager reduce the price of an acre-foot of water by $18.37 in the same year that he mandates the adoption of low-flow appliances as part of a conservation program. The same municipal manager, however, when faced with the need to start gradually raising prices 15 years later (as he is in Figure 1) might use the added flexibility allowed by the low-flow appliance adoptions to not do so *as rapidly* (i.e., yielding a gradual decline in the unconstrained low-low adoption price path relative to the status quo price path). While the lost conservation potential of the low-flow devices might not therefore be as large as the 80% described in Section 5, it is likely to be significant. This corresponds with the anecdotal evidence received on mandatory low-flow appliance adoptions – they are most successful when adopted in conjunction with a program of pricing austerity.

With a policy objective of achieving a reduction in lift-height equal to that implied by mandatory low-flow appliance adoption with unconstrained pricing, is there a socially preferable way to achieve the result? Unless consumer surplus is not reduced at all by the introduction of low-flow appliances (i.e., unless the exact same consumer surplus is achieved in spite of the fact that less water is used), a small water tax will be superior in terms of social surplus (i.e., aggregating the welfare consequences for households across their roles as water consumers and taxpayers).

Given the anecdotal evidence described in Section 2, it is unlikely that consumer surplus does not suffer under mandatory low-flow appliance installation (holding prices constant), making a modest water tax the cost-effective policy. Why then do rational water managers so frequently choose a cost-ineffective policy? The answer most probably lies in equity considerations. Water is a necessity, while most forms of municipal revenue collection (used to offset water utility deficits) are progressive. Structural parameter estimates imply that municipal managers weight more heavily the welfare of water consumers relative to that of taxpayers, suggesting a generally progressive water provision-revenue collection platform.

In any case, the evidence is clear that, if the goal of demand-side technology standards is water conservation and not income redistribution, they must be accompanied by pricing austerity – status quo pricing in the least – to be effective. If income redistribution is the goal of policy makers, more direct and efficient fiscal tools should be employed.

Appendix: Dynamic Estimation Algorithm

A nested maximum likelihood estimation procedure is used to recover the parameters underlying the municipal manager's dynamic payoff-maximization problem. That procedure uses a continuous state-continuous (but constrained) control version of a numerical fixed-point algorithm to find the solution to the manager's dynamic programming problem and employs that solution in deriving the likelihood function. Finally, those likelihood function evaluations are nested in a quasi-Newton hill-climbing algorithm. Owing to the high incidence of constrained pricing decisions observed in the data (in particular, municipal managers charge a marginal price of zero in nearly half of all observations, see Table 1), more traditional Euler equation techniques, based upon simple offsetting variations, cannot be used without throwing-out nearly half of the data.

The following discussion describes the derivation of the conditional likelihood function that is used to estimate the parameters of the municipal manager's dynamic objective function. The full likelihood function measures the probability of realizing an observed vector of data, $[P_{i,t}, h_{i,t}, \mathbf{X}_{i,t}, D_{i,t}]$, as well as a vector of unobserved net marginal revenues, $\xi_{i,t}$, as a function of a parameter vector, $\mathbf{\Theta}$. $\mathbf{X}_{i,t}$ is presumed to evolve according to a process determined by the parameter vector, χ, which is estimated independently and used only in the counterfactual simulations conducted in Section 5. Temporarily treating $\xi_{i,t}$ as if it were observed data, the probability density in which we are interested can be written in its most primitive form as $f(\mathbf{h}, \mathbf{P}, \mathbf{D}, \boldsymbol{\xi}, \mathbf{X}; \mathbf{\Theta})$. Conditioning upon the exogenous states, $\mathbf{X}$, and assuming that the conditional joint density of $\mathbf{P}, \mathbf{h}, \mathbf{D}$, and $\boldsymbol{\xi}$ $\forall i, t$ is independent across cities, this density can be rewritten as:

$$f(\mathbf{h}, \mathbf{P}, \mathbf{D}, \boldsymbol{\xi} \mid \mathbf{X}; \boldsymbol{\Theta}) = \prod_{i=1}^{13} f_i(\mathbf{h}_i, \mathbf{P}_i, \mathbf{D}_i, \boldsymbol{\xi}_i \mid \mathbf{X}_i; \boldsymbol{\Theta}). \tag{A.1}$$

Next, recall from Equation (4.1) that $D_{i,t}$ is a function of only $P_{i,t}$, $\mathbf{X}_{i,t}$, a stochastic term, and $\boldsymbol{\Theta}$. Equation (4.5) defines $h_{i,t}$ as a function of only $h_{i,t-1}$, $D_{i,t-1}$, $\mathbf{X}_{i,t-1}$, $\varepsilon^h_{i,t-1}$, and $\boldsymbol{\Theta}$ (including $Z_{i,t-1}$ as an element of $\mathbf{X}_{i,t-1}$ for simplicity), and (4.4) defines $\xi_{i,t}$ as a function of only $\xi_{i,t-1}$, $\varepsilon^\xi_{i,t-1}$, and $\boldsymbol{\Theta}$. Finally, note that the dynamic programming problem described in Equation (4.6) yields a conditional policy function that defines price, $P_{i,t}$, as a function of only $h_{i,t}$, $\mathbf{X}_{i,t}$, $\xi_{i,t}$, and $\boldsymbol{\Theta}$. With these four results, each term on the right-hand side of (A.1) can be factored according to an iterative procedure,[14] yielding:

$$\begin{aligned} f(\mathbf{h}_i, \mathbf{P}_i, \mathbf{D}_i, \boldsymbol{\xi}_i \mid \mathbf{X}_i; \boldsymbol{\Theta}) \\ = \Bigg[\prod_{t=1}^{T_i-1} & f_D(D_{i,t+1} \mid P_{i,t+1}, \mathbf{X}_{i,t+1}; \boldsymbol{\Theta}) \times f_p(P_{i,t+1} \mid h_{i,t+1}, \mathbf{X}_{i,t+1}, \xi_{it}; \boldsymbol{\Theta}) \\ & \times f_h(h_{i,t+1} \mid h_{i,t}, D_{i,t}, \mathbf{X}_{i,t}; \boldsymbol{\Theta}) \times f_\xi(\xi_{i,t} \mid \xi_{i,t-1}; \boldsymbol{\Theta})\Bigg] \\ & \times f(P_{i,1}, h_{i,1}, D_{i,1} \mid \mathbf{X}_{i,1}, \mathbf{X}_{i,0}; \boldsymbol{\Theta}), \end{aligned} \tag{A.2}$$

where $\xi_{i,1}$ is distributed conditional upon $\xi_{i,0}$, a parameter to be estimated for each city.

Each of the probabilities in this expression can be calculated as follows. First, given the distributional assumptions outlined in Section 4, conditional probability densities for D and h are calculated by a simple change-of-variables. The specification of the dynamic programming problem in (4.6) and the invertibility of the policy function over $\xi_{i,t}$, given $\xi_{i,t-1}$ and $\boldsymbol{\Theta}$, suggest an iterative process for deriving the conditional probabilities of P and ξ. In particular, for each realization of $\mathbf{X}_{i,t}$, the conditional policy function, $P(h_{i,t}, \xi_{i,t} \mid \mathbf{X}_{i,t}; \boldsymbol{\Theta}^1)$, is calculated at every point in $[h, \xi]$ space via a combination of value and policy function iteration for some initial parameter guess $\boldsymbol{\Theta}^1$. In every case, the conditional value function is monotonically decreasing in lift-height and monotonically increasing in unobserved net marginal revenues. The conditional policy function, conversely, is monotonically increasing in lift-height and weakly monotonically decreasing in unobserved net marginal revenues. Next, using the distribution ascribed to the unobserved component of net marginal revenues, $\xi_{i,t}$, and controlling for the observed value of $h_{i,t}$, the conditional policy function is inverted to arrive at a conditional probability density for $P_{i,t}$:

$$f_P(P_{i,t} \mid h_{i,t}, \mathbf{X}_{i,t}; \boldsymbol{\Theta}^1) = [f_{P|P=0}]^{\lambda_{i,t}}[f_{P|P>0}]^{1-\lambda_{i,t}}, \tag{A.3}$$

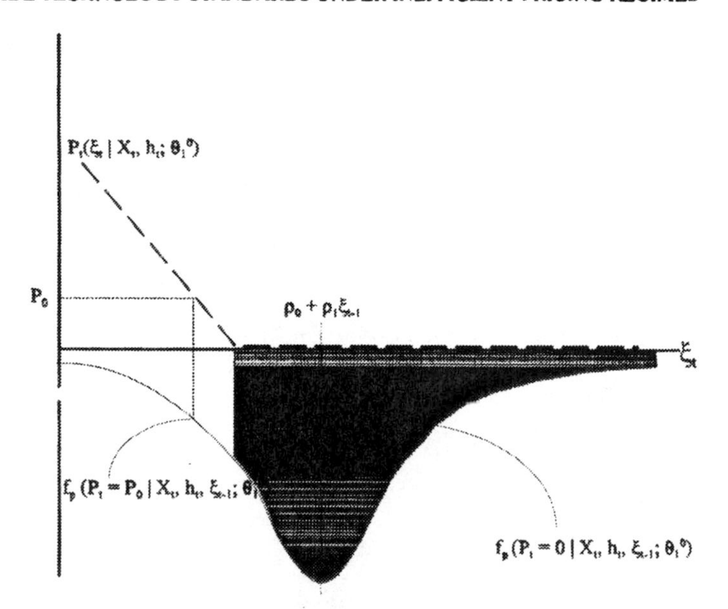

Figure 3. Condition policy function.

where $\lambda_{i,t} = 1$ if $P_{i,t} = 0$, $\lambda_{i,t} = 0$ if $P_{i,t} > 0$, and

$$f_{P|P=0} = 1 - \Phi\left(\frac{\xi^*_{i,t} - \mu_\xi}{\sigma_\xi}\right), \quad f_{P|P>0} = \frac{\varphi\left(\frac{g^{-1}(P_{i,t}) - \mu_\xi}{\sigma_\xi}\right)}{\Phi\left(\frac{\xi^*_{i,t} - \mu_\xi}{\sigma_\xi}\right)} \times \left\| \frac{\partial g^{-1}}{\partial P_{i,t}} \right\| \quad \text{(A.4)}$$

(see Figure 3). $\xi^*_{i,t}$ denotes the value of ξ where, conditional upon observed $h_{i,t}$, the non-negativity constraint on price begins to bind; i.e., $P_{i,t} = g(\xi_{i,t})$ for all $\xi_{i,t} < \xi^*_{i,t}$. Finally, given the conditional policy function, conditioning upon observed $h_{i,t}$ and the parameter vector Θ^1, and given some $\xi_{i,t-1}$ calculated in the previous iteration (or given the parameter $\xi_{i,0}$ if in the first iteration), observed $P_{i,t} > 0$ determines $\xi_{i,t}$. This value of $\xi_{i,t}$ then becomes the basis for the calculation of the probability of observing $P_{i,t+1}$ in the next iteration, given Θ^1 and observed $h_{i,t+1}$.

For cities charging $P_{i,t} = 0$ (recall that this applies to approximately one-half of the sample), calculating the likelihood function is more difficult. In particular, a unique value of $\xi_{i,t}$ cannot be recovered by inverting the policy function; i.e., the only available information is that $\xi_{i,t}$ lies in the flat region of the policy function. Instead, the expectation of the entire likelihood expression for these cities must be calculated with respect to the distribution of the Markov chain of ξ's. Specifically, for the six cities in the panel that always charge a marginal price of zero, a Monte Carlo simulation approach is used to calculate their expected contribution to the likelihood function. First, conditional policy functions are calculated for each of these cities in every year, over every possible value of ξ. A sequence of ξ's are then simulated for each of these cities, conditional upon $P_{i,t} = 0$ given the conditional policy function (i.e., random draws determining the increment to ξ each period come from truncated normal distributions, with the truncation point being determined in period t by $\xi_{i,t-1}$ and the value of $\xi^*_{i,t}$ where the policy function becomes constrained in period t). The likelihood function is then calculated for each city for each simulated sequence of ξ's. $f_p(P_{i,t} \mid h_{i,t}, \mathbf{X}_{i,t}, \xi_{i,t-1}; \boldsymbol{\Theta}^1)$ is still given by the integral expression in (A.4) when $P = 0$, but $\xi_{i,t-1}$, which enters that expression through μ_ξ, comes from a simulation draw of $\varepsilon^\xi_{i,t}$ and the previously simulated value of $\xi_{i,t-2}$. The likelihood is calculated for each of 1,000 simulated sequences of ξ's, making simulation error negligible. The expectation of city i's contribution to the likelihood function is found by taking the average of these 1,000 simulated likelihoods.

The conditional likelihood of observing all the data given the parameter vector $[\boldsymbol{\Theta}^1, \boldsymbol{\xi}^1_{,0}]$ can be written:

$$
\begin{aligned}
&L(\mathbf{h}, \mathbf{P}, \mathbf{D}, \boldsymbol{\xi} \mid \mathbf{X}, \mathbf{P}_1, \mathbf{h}_1; \boldsymbol{\Theta}^1, \xi^1_0) \\
&= \Bigg[\prod_{p>0} \prod_{t=1}^{T_i - 1} f_D(D_{i,t+1} \mid P_{i,t+1}, \mathbf{X}_{i,t+1}; \boldsymbol{\Theta}^1) \\
&\times f_P(P_{i,t+1} \mid h_{i,t+1}, \mathbf{X}_{i,t+1}, \xi_{i,t}; \boldsymbol{\Theta}^1) \\
&\times f_h(h_{i,t+1} \mid h_{i,t}, D_{i,t}, \mathbf{X}_{i,t}; \boldsymbol{\Theta}^1) \times f_\xi(\xi_{i,t} \mid \xi_{i,t-1}; \boldsymbol{\Theta}^1) \Bigg] \\
&\times \prod_{P=0} E \Bigg[\prod_{t=1}^{T_i - 1} f_D(D_{i,t+1} \mid P_{i,t+1}, \mathbf{X}_{i,t+1}; \boldsymbol{\Theta}^1) \\
&\times f_P(P_{i,t+1} \mid h_{i,t+1}, \mathbf{X}_{i,t+1}, \boldsymbol{\xi}_{i,t}; \boldsymbol{\Theta}^1) \\
&\times f_h(h_{i,t+1} \mid h_{i,t}, D_{i,t}, \mathbf{X}_{i,t}; \boldsymbol{\Theta}^1) \Bigg].
\end{aligned} \tag{A.5}
$$

New values of $\boldsymbol{\Theta}^2$ and $\boldsymbol{\xi}^2_0$ are determined so as to increase L, and estimates of $\boldsymbol{\Theta}$ and $\boldsymbol{\xi}_0$ are those values that maximize this expression. These estimates are both consistent and asymptotically efficient as $T \to \infty$.[15]

This algorithm would be prohibitively time-consuming given the 68 parameters that comprise $\boldsymbol{\Theta}$. As such, a two-step estimation routine is used, whereby all parameters in $\boldsymbol{\Theta}$ except ν, ρ_0, ρ_1, σ_ξ, and $\boldsymbol{\xi}_0$ are estimated using the full data panel in a first-stage instrumental variables procedure that does not require the solution to a dynamic program. These estimated parameter values are then used in the dynamic maximum likelihood estimation of the remaining parameters as if they were known with certainty. Observations not satisfying dynamic data requirements are dropped in this second stage, leaving a total of 116 observations. The standard errors of the second-stage estimates are then calculated while accounting for the error introduced by the first-stage parameter estimates.

Notes

[1] Water is not the only commodity where demand-side technology standards are used for conservation. While the issue of short-run exhaustion of reserves is not as significant as in the case of water, numerous demand-side technology adoption programs have been established to reduce energy consumption and the depletion of non-renewable fossil fuel stocks. The Oregon Residential Energy Tax Credit Program, for example, provides tax credits for energy-efficient appliances, premium efficiency ducts, geothermal space heating, solar water heating, solar space heating, solar electricity, and alternative-fuel vehicles (www.energy.state.or.us/res/tax/taxcdt.htm). Programs are in place in New Jersey (www.southjerseygas.com/appliances/fin2000.htm) and Mesa, AZ (www.ci.mesa.az.us/utilitie/gas/gasreb.htm) to support conversion to high-efficiency natural gas heaters and appliances, while rebates for energy-efficient home cooling systems have been used in Austin, TX (www.austinenergy.com/home/coolers-e.html). Similarly, in Boston (www.easterngas.com/ideas/body/rebate.htm), Hawaii (www.heco.com/energyservices/), and Indiana (www.citizensgas.com/residential/rebates.htm), rebates have been given to homeowners for switching to high-efficiency natural gas water heaters. Residential *Energy Star* programs have provided rebates to homeowners for the installation of low energy-use refrigerators, air conditioners, dish-washers, light-bulbs, or clothes-washers in Pasadena, CA (www.ci.pasadena.ca.us/waterandpower/estar.asp), Connecticut (www.uinet.com/yourhome/Energyindex.asp), and Eugene, OR (www.eweb.org/energy/energysharp). Similar programs have been instituted for businesses in Oregon (www.energy.state.or.us/bus/tax/taxcdt.htm), Wisconsin (www.nspco.com/fb/fb_ps_wi.htm), Minnesota (www.nspco.com/fb/fb_ps.htm and www.austinutilities.com/commercial/energyef.htm), Boston (www.easterngas.com/ideas/body/rebatebus.htm), New York (www.nrdc.org/cities/building/nnytax.asp), and California (www.pge.com/customer_services/ business/energy/express/). Instead of direct subsidies, some programs have provided subsidized loans to businesses (www.nol.org/home/NEO/loan/improv.htm) and subsidized mortgages to homeowners (www.sdge.com/res/homeloan.html).

[2] *The Denver Post* (Martin, 1999) reports that, in a recent survey, 75% of homeowners and 78% of property managers who had installed low-flow toilets had experienced problems of these sorts in the last year. With an office located next to a bathroom with a retrofitted low-flow toilet in a 134-year-old building, this author can personally attest to these kinds of problems as well. The cost of plumbing and household repair expenditures (i.e., up to $100 per household) are typically ignored when measuring the costs of water conservation with demand-side technology standards.

[3] An 18' × 36' foot pool can evaporate 20,000 gallons of water a year in the hot southern California climate.

[4] This income measure is adjusted to account for the fixed charges each household pays in order to connect to the water provision system and any other inframarginal charges above or below the marginal price. This adjusted measure is typically referred to as 'virtual income' (Nordin, 1976). All price and income figures in the data are deflated to constant 1982–1984 dollars.

[5] Here, and in the remainder of the chapter, boldface is used to indicate a vector. When describing the data panel used for estimation, $\mathbf{X}$ refers to a vector over both time (t) and cities (i). $\mathbf{X}_i$ refers to a vector over only the time dimension for city i.

[6] Lift-height measures the distance, in feet, that water needs to be pumped to the surface. An increase in lift-height therefore measures a decrease in the available stock of groundwater.

[7] Φ_0 includes $\mathbf{X}_0$ and h_0, but not ε_0^d, ε_0^c, ε_0^S, or ε_0^X.

[8] An acre-foot of water is the volume of water needed to cover one acre at a depth of one foot. This standard unit of measurement in water provision is equivalent to 43,560 cubic feet or 325,851 gallons.

[9] Assuming that population and real income rise gradually over time, this assumption implies that the modeled municipal manager would consistently under-predict future water demand. Since the manager cares about being able to supply water to future customers at a reasonable price, this means that he will consistently underprice water in the current period. In order to get the model's price predictions to correspond to observed pricing behavior, this assumption will therefore bias down the estimate of ν (i.e., in the direction of more weight being placed on taxpayer welfare). Since the model seeks to determine whether ν *exceeds* 1/2 by a significant amount (in order to explain observed underpricing decisions), this bias works against the alternative hypothesis, $H_A : \nu > 1/2$. This makes the statistical evidence of $\nu > 1/2$ that we recover below even stronger. Finally, since the assumption of myopic manager beliefs about $\mathbf{X}_t$ biases ν closer to efficiency (i.e., $\nu = 1/2$), it should also bias down the size of the deadweight efficiency losses simulated in the following section.

In practice, similar empirical results were obtained in an estimation model that assumed that managers had perfect foresight about $\mathbf{X}_{t+1}$ in period t, so it is unlikely that the myopia assumption has a significant impact on the results.

[10] Non-uniform accounting standards across municipalities imply that expenditures for the same non-pumping activity in different cities might not be included in any reported cost-category. The empirical model accounts for this source of measurement error with the introduction of the structural error term, $\xi_{i,t}$, which measures unreported net revenues per unit.

[11] This approach fails, however, to account for strategic interactions between neighbors. Allowing for such interactions would necessitate the introduction of a Markov Perfect Nash Equilibrium, making estimation computationally impractical.

[12] In practice, a two-step estimation routine was employed in order to obtain improved parameter estimates of coefficients that did not depend upon model dynamics for identification. This is described in Timmins (2002) and the Appendix.

[13] Estimates suggest that approximately 30% of indoor household indoor water is used in the shower, while another 30% is used in the toilet. This policy approximates a 50% reduction in either of these sources of water use, such as that which would be achieved by switching from a 3.5 gallon toilet to an ultra-low-flow (1.6 gallon) toilet. One might think of this policy as reducing the quantity of water needed to take a 'typical' shower by 50%. Changes in price of water, however, can still alter the length of shower chosen by the homeowner.

[14] First, the conditional density of demand in the final year for city i is separated from the joint density of the remainder of city i's data. The conditional density of city i's price in that year is then separated from the remaining joint density. A similar process is conducted for lift-height in year T, and ξ in year $T-1$. The resulting univariate conditional probabilities are simplified with the functional forms described in (4.1), (4.4), and (4.5), and the process is repeated for year $T-1$. Repeating this procedure $T-1$ times for each city, and conditioning on the joint distribution of $P_{i,1}$, $h_{i,1}$, and $D_{i,1}$ and on the estimated parameters $\xi_{i,0}$ yields the expression in (A.2).

[15] The term $\Pi_i f_i(P_{i,1}, h_{i,1}, D_{i,1} \mid \mathbf{X}_{i,1}, \mathbf{X}_{i,0}; \mathbf{\Theta})$, upon which this likelihood function is conditioned, becomes inconsequential in identifying $\mathbf{\Theta}$ and $\xi_{i,0}$ as the data grows in the time dimension. Moreover, the term $f_X(\mathbf{X}; \chi)$ has no bearing on the estimation of $\mathbf{\Theta}$ and $\xi_{i,0}$. $f_X(\mathbf{X}; \chi)$ can thus be excluded from the conditional likelihood function as well.

References

Austin Energy Homepage, Internet, available 21 October 2000, www.austinenergy.com/home/coolers-e.html.

Austin Utilities Commercial and Industrial Customer Programs Homepage, Internet, available 21 October 2000, www.austinutilities.com/commercial/energyef.htm.

Bernstein, S. (1996) Drought's over, but conservation habits still reign, *The Los Angeles Times*, 28 May, p. B1.

Booth, M. (1999) Tyranny of toilet police destined for a flushing, *The Denver Post*, 30 September, p. B2.

Bostongas Homepage, Internet, available 21 October 2000, www.easterngas.com/ideas/body/rebate.htm and www.easterngas.com/ideas/body/rebatebus.htm.

California Department of Finance (1970–1993) Population estimates of California cities and counties, Reports (70 E-1)–(93 E-1), Department of Finance, Population Research Unit, Sacramenta, CA.

California Department of Water Resources (1992) Historical unconfined ground water trends in the San Joaquin Valley, San Joaquin District Office, Sacramenta, CA.

California Department of Water Resources (1993) *The California Water Plan Update*, Bulletin 160-93, Sacramenta, CA.

Citizens Gas Homepage, Internet, available 21 October 2000, www.citizensgas.com/residential/rebates.htm.

City of Mesa, Utilities Department Homepage, Internet, available 21 October 2000, www.ci.mesa.az.us/utilitie/gas/gasreb.htm.

The Columbus Dispatch (1999) Commode-tion humor aside, flap over flushes no joke, 4 August, p. A10.

Eugene Water and Electric Board Homepage, Internet, available 21 October 2000, www.eweb.org/energy/energysharp.

Frammolino, R. (1992) High-volume debate over low-flows, *The Los Angeles Times*, 23 February, p. B1.

Hawaiian Electric Company, Inc. Homepage, Internet, available 21 October 2000, www.heco.com/energyservices.

Johnson, S. (1999) Low-flow critics want feds out of bathrooms, *The Arizona Republic*, 19 September, p. A24.

La Rue, S. (1995) Residents snap up low-flow toilet rebates, *The San Diego Union-Tribune*, 13 February, p. B8.

Lilly, D. (1992) Water rules end, rates head up – 20 percent increase on tap in '93, *The Seattle Times*, 11 March, p. C1.

The Los Angeles Times (1991) San Gabriel Valley and the drought; Rebates offered for conservation, 21 March, p. J6.

Martin, K. (1999) Water-stingy toilets costly to maintain, *The Denver Post*, 19 September, p. A39.

Natural Resources Defense Council Homepage, Internet, available 21 October 2000, www.nrdc.org/cities/building/nnytax.asp.

Nebraska Energy Office Homepage, Internet, available 21 October 2000, www.nol.org/home/NEO/loan/improv.htm.

Nordin, J.A. (1976) A proposed modification of Taylor's demand analysis: Comment, *The Bell Journal of Economics* **7**, 719–721.

Northern States Power Company Homepage, Internet, available 21 October 2000, www.nspco.com/fb/fb_ps_wi.htm and www.nspco.com/fb/fb_ps.htm.

Oregon Residential Energy Tax Credit Program Home Page, Internet, available 21 October 2000, www.energy.state.or.us/res/tax/taxcdt.htm and www.energy.state.or.us/bus/tax/taxcdt.htm.

Pacific Gas and Electric Homepage, Internet, available 21 October 2000, www.pge.com/customer_services/business/energy/express.

Pasadena Water and Power Homepage, Internet, available 21 October 2000, www.ci.pasadena.ca.us/waterandpower.estar.asp.

Peltzman, S. (1971) Pricing in public and private enterprises: Electric utilities in the United States, *The Journal of Law and Economics* **14**, 109–147.

San Diego Gas and Electric Homepage, Internet, available 21 October 2000, www.sdge.com/res/homeloan.html.

Socratous, G. (2000) Water policy issues in Cyprus, Mimeo, Symposium on Water Resources Management, Efficiency, Equity and Policy, University of Cyprus, Nicosia.

South Jersey Gas, Gas Advantage Stores Home Page, Internet, available 21 October 2000, www.southjerseygas.com/appliances/fin2000.htm.

Stokey, N.L. and Lucas Jr., R.E. (1989) *Recursive Methods in Economic Dynamics*, Harvard University Press, Cambridge, MA.

Timmins, C. (2000) Does the median voter consume too much water? Analyzing the redistributive role of residential water bills, Mimeo, Yale University Department of Economics.

Timmins, C. (2002) Measuring the dynamic inefficiency costs of regulators preferences: Municipal water utilities in the arid West, *Econometrica*, forthcoming.

The United Illuminating Company Homepage, Internet, www.uinet.com/yourhome/Energyindex.asp, available 21 October 2000.

PART 4: WATER MANAGEMENT POLICIES

4.1. *Water as a Multifaceted Resource – Horizontal Management Considerations*

Joint Quantity/Quality Management of Groundwater

CATARINA ROSETA PALMA*

1. Introduction

Aquifers have always been used as sources of water, and their importance in water management can be great. It is widely known that water as a resource needs to be managed beyond mere quantity allocations, since the quality of the water is often decisive when determining how much of it will be used. Given the dynamic nature of groundwater, this means both the stock of water held by an aquifer and the stocks of whatever pollutants contaminate it are essential variables.[1] Many existing aquifers, especially those in semi-arid regions where agricultural production is intense, present both scarcity problems (overdraft) and quality deterioration. In Portugal, for example, such is the case with some important aquifers in the South (Évora, Beja, Algarve).

Groundwater is usually exploited in a common property regime, in the sense that access is limited to owners of the land overlying the aquifers. This creates externalities since one unit of water extracted is no longer available to others and lowers the water table for everyone. Additional externalities are present because individual contamination affects all users. Furthermore, there are several reasons why the relationship between water pumping decisions and contamination cannot be ignored. First, the economic decision to pump is associated with the choice of polluting inputs. Second, the actual contamination that percolates to the aquifer also depends on how much water is applied on the surface. And finally, the quality of the water may alter its productivity. Yet the bulk of existing economic literature treats these two aspects of groundwater management independently.

Evidently, common property solutions are always associated with lower levels of social welfare than what could be achieved under optimal management. Two interesting questions arise then: how big is this welfare loss and what can be done about it? The answer to the first question should allow decision makers to decide whether or not to intervene in the aquifer; if intervention is desirable, the answer to the second question should allow them to decide on the most appropriate type of policy instrument to apply. The size of the welfare loss brought about by common property arrangements is largely an empirical question. In water quantity management, a few studies exist, beginning with Gisser and Sanchez (1980), although

* Financial support from the Fundação Amélia de Mello is gratefully acknowledged.

P. Pashardes et al. (eds.),
Current Issues in the Economics of Water Resource Management, 151–170.

results vary somewhat (see Provencher, 1995, for a survey). However, for quality management, perhaps due to the negative impacts on public health often associated with contamination of aquifers, or because remediation programs tend to be difficult and expensive, the need for intervention is no longer an issue. Most countries already have maximum standards established for the most important pollutants (for instance, the US Clean Water Act, the EU Nitrates Directive EC 91/676) Although these requirements are not necessarily those that maximize welfare, their existence establishes that intervention is warranted. So here the choice of policy instruments is the key issue.

In the quantitative management literature, policy recommendations include water charges and water quotas which essentially lead users to reproduce the pumping paths of central control (see Feinermann and Knapp, 1983; Neher, 1990). A less interventionist strategy is described by Provencher (1993), where users receive an endowment of tradable permits which they control over time. Typical pollution control literature also suggests variations on taxes or permits. Instruments can be targeted at actual discharges or at the inputs that cause them, which are easier to monitor and will be equally efficient as long as the pollution production function is known (see Griffin and Bromley, 1982). Since complete information on pollution processes is frequently unavailable, incentives can take the form of penalties on deviations between observed and desirable discharges (Segerson, 1988). Xepapadeas (1992) presents such incentives for the case of a dynamic pollutant accumulation model with imperfect monitoring. Given implementation difficulties, there is also some work in the literature evaluating suboptimal pollution control instruments (see Shortle and Dunn, 1986; Larson et al., 1996; Helfand and House, 1995; Mapp et al., 1994).

This chapter uses a dynamic resource management perspective to show how an optimal policy takes into account effects on water stock as well as on its quality. The cases in which it does not are shown to be special cases of the joint management solution. In a model where an output is produced using water and another input, which then combine to cause pollution, the optimal taxes on groundwater extraction and on the polluting input are calculated assuming that the producers/users of the water are myopic. It is shown that taxes based on typical models of quantity-only or quality-only are inefficient and may not even lead the common property aquifer in an optimal direction. Similar results are obtained when quality enters the problem as a restriction instead of directly affecting productivity. Implementation problems associated with optimal policies are briefly discussed, and an illustration is provided.

2. Why Private Pumpers Do Not Maximize Welfare

The exploitation of a stock of groundwater is typically a problem of common property, since there is limited access to the resource. If each user is a profit maximizing agent, there are a number of well known reasons for the inefficiency of the private solution.

First, the fact that there is a finite stock means that each unit that is extracted by one firm is no longer available to the others. Given that the only way to lay claim to a unit of groundwater is to pump it, there is little incentive to save water for later use, since other firms have the same access to the stock. Hence firms will pump an inefficiently large amount of groundwater. Moreover, pumping costs generally depend on the water table level, which means that as the groundwater stock is depleted, extraction costs rise. However, the individual user does not consider the detrimental effect of its pumping on other firms' costs, so again there is an incentive to overpump.

Finally, if the aquifer under consideration is susceptible of contamination resulting from users' actions in the land overlying the aquifer, then additional externalities occur. Since each user will only consider his own polluting impact on the stock (or not even that one, if the agent behaves myopically), he has a motive to overpollute, imposing external costs on all other users of the water. Note that the presence of contamination implies that private agents will implement not only a suboptimal choice for water but also a suboptimal choice for contaminating actions (such as fertilizer or pesticide use). There is also less incentive to invest in groundwater protection and/or remediation programs.

Considering the presence of uncertainty would originate yet another externality. In a conjunctive system where surface water flows are stochastic, groundwater plays a stabilizing role by reducing the variability of available water and thus that of users income. A large stock protects all users from income riskiness, so that each unit of water left in the ground has risk reducing value. When making extraction decisions, however, each user fails to consider the buffer role of leftover units, and his extraction path is thus suboptimal (see Provencher and Burt, 1993).

3. Joint Quantity/Quality Model

Consider a dynamic, continuous time model of groundwater management for an aquifer with constant recharge, R. Assume that M identical firms exploit a single stock of groundwater, which contains G_t units of recoverable water and is characterized by a flat bottom and perpendicular sides. The water has a known level of quality, Ψ, which affects the profitability of water use. Worsening quality might be due to nitrates in the water, dissolved solids, trace metals, or any other relevant quality measure. All firms produce some good using water and another input with a production function $y(g_t, \gamma_t; \Psi_t)$, where g_t is the amount of water used and γ_t is an input whose use creates pollution. The following properties are assumed:

- positive but diminishing marginal returns for both inputs: $\partial y/\partial g, \partial y/\partial \gamma > 0$, $\partial^2 y/\partial g^2, \partial^2 y/\partial \gamma^2 < 0$;
- complementarity of inputs: $\partial^2 y/\partial g \partial \gamma > 0$;
- positive effect of water quality on production:[2] $\partial y/\partial \Psi > 0$;
- positive effect of water quality on marginal productivities of other inputs: $\partial^2 y/\partial g \partial \Psi > 0, \partial^2 y/\partial \gamma \partial \Psi > 0$.

The unit cost of groundwater extraction, denoted by $c(G_t)$ is decreasing and convex). Labelling each firm's extraction of groundwater g_t, each firm's net benefit at period t is given by:

$$p_y y(g_t, \gamma_t; \Psi_t) - c(G_t) g_t - p_\gamma \gamma. \tag{1}$$

Considering a return coefficient of α, the groundwater stock behaves according to:

$$\frac{\partial G_t}{\partial t} \equiv \dot{G} = -M(1-\alpha) g_t + R. \tag{2}$$

Water quality will degrade as it receives contaminant loads, which originate on the surface and percolate towards groundwater according to a pollution production function that depends on how much water is used and on the amount of polluting input, $e(g_t, \gamma_t)$, so that $\partial e/\partial g_t > 0$, $\partial e/\partial \gamma_t > 0$ and $\partial^2 e/\partial g_t \partial \gamma_t > 0$. Increasing amounts of water are assumed to increase pollution because water is a carrier for the pollutants. However, this effect could be negative, at least for some range of w, if other roles of water were considered, namely its complementarity with polluting inputs in the production of y (for instance, in agriculture, water allows plants to absorb nitrogen and phosphorus more efficiently decreasing the presence of these elements in topsoil) and its own dilution effect (if all the contaminant load is already being carried down, the addition of extra water will improve the quality of the leachates that reach the aquifer). Examples of pollution production functions for nitrates in agriculture can be found in Vickner et al. (1998); Larson et al. (1996) and Helfand and House (1995).

The evolution of groundwater quality depends on how much pollution the aquifer receives and on its ability to regenerate. Generally the quality of the recharge will also affect Ψ, but this effect is exogenous and it will be ignored to simplify the model. Thus:

$$\dot{\Psi} = f(\Psi_t, e(g_t, \gamma_t)). \tag{3}$$

Following the contamination literature (see Yadav, 1997, for nitrates; Anderson et al., 1985, for pesticides), it is assumed that there is a constant natural decay rate for the pollutant, so that contamination would evolve according to:

$$\dot{C} = Me(g_t, \gamma_t) - \delta C_t. \tag{4}$$

A more realistic version of this equation would include delayed response effects, as pollutants do not reach groundwater instantly. Note that if stock effects existed, they could enter the function through their impact on the decay rate, which might be more generally represented as $\delta(C, G)$.

The quality measure should be defined so that it decreases with contamination. One possibility is to define quality as the difference between the maximum possible stock of contaminant the aquifer can take and its actual presence, $\Psi_t = C^{\max} - C_t$, implying that the associated quality function is:[3]

$$\dot{\Psi} = \delta(\Psi^{\max} - \Psi_t) - Me(g_t, \gamma_t). \tag{5}$$

3.1. Optimal vs. Common Property Management

The problem facing a water planner would be to choose optimal extraction paths and input use for each firm, i.e. to determine which paths maximize the total present value of net revenues. It is assumed that there are no environmental externalities, otherwise these would have to be considered as well.

Formally, the water planner's problem is:

$$\max_{g_t} \int_0^\infty M\left[p_y y(g_t, \gamma_t; \Psi_t) - c(G_t)g_t - p_\gamma \gamma\right] e^{-\rho t}\, dt, \tag{6}$$

subject to Equations (2) and (5) and to nonnegativity restrictions and/or maximum levels for G_t and Ψ_t, as well as initial conditions $G_0 = \overline{G}$ and $\Psi_0 = \overline{\Psi}$. ρ is the appropriate discount rate, which is assumed to be constant. This is a typical optimal control problem. The current value Hamiltonian is:

$$\begin{aligned} H &= M\left(p_y y(g_t, \gamma_t; \Psi_t) - c(G_t)g_t - p_\gamma \gamma\right) \\ &\quad + \lambda_t(-M(1-\alpha)g_t + R) + \beta_t\left(\delta(\Psi^{\max} - \Psi) - Me(g_t, \gamma_t)\right). \end{aligned} \tag{7}$$

The results obtained from first order conditions can be summarized as:[4]

$$p_y \frac{\partial y}{\partial g_t} = c(G_t) + \lambda_t(1-\alpha) + \beta_t \frac{\partial e}{\partial g_t}, \tag{8}$$

$$p_y \frac{\partial y}{\partial \gamma_t} = p_\gamma + \beta_t \frac{\partial e}{\partial \gamma_t}, \tag{9}$$

$$\dot{\lambda} = \rho \lambda_t + M \frac{\partial c}{\partial G_t} g_t, \tag{10}$$

$$\dot{\beta} = (\rho + \delta)\beta_t - Mp_y \frac{\partial y}{\partial \Psi_t}. \tag{11}$$

The first equation represents the usual optimality result that marginal benefit in each period will be equal to total marginal extraction cost, which is the sum of three terms: actual extraction cost, opportunity cost of removing one unit of water from the ground (reflecting the future impact on profits for all firms), and an additional cost due to the impact of extraction water, which is applied then returns with contaminants, on water quality. The second equation shows an equivalent result for the other input: marginal benefit is equal to actual marginal cost plus the cost of quality deterioration. The third and fourth equations describe the behaviour of the shadow prices of quantity and quality, respectively. Note that the evolution of β reflects the positive impact of quality on profits as well as the regeneration rate.

In the presence of a positive recharge and a natural regeneration rate, it is possible to have a steady state with positive pumping and positive discharges, for which: $\dot{G} = 0$, $\dot{\Psi} = 0$, $\dot{\lambda} = 0$ and $\dot{\beta} = 0$. These yield, respectively (variables are at their steady state values):

$$g = \frac{R}{(1-\alpha)M}, \tag{12}$$

$$\Psi = \Psi^{\max} - \frac{Me(g,\gamma)}{\delta}, \tag{13}$$

$$\lambda = \frac{-M\frac{\partial c}{\partial G}g}{\rho}, \tag{14}$$

$$\beta = \frac{Mp_y\frac{\partial y}{\partial \Psi}}{\rho + \delta}. \tag{15}$$

From these equations it can be noted that at the steady state pumping is exogenous (Equation (12)), quality will be higher for lower levels of pollution (Equation (13)), the shadow price of water in the ground will be higher for lower levels of stock (Equation (14)), and the shadow price of water quality will be higher when the impact of quality on production is higher (Equation (15)). Note that in the special case where unit cost is approximately constant, $\partial c/\partial G = 0$, and the constraint $g_t \leq G_t$ is not binding, which might happen in very large aquifers with sizable recharges (implying a relatively stable water stock), a quality-only model could be appropriate, while if $\partial y/\partial \Psi = 0$ a quantity-only model is appropriate.

In a common property situation, if the users of groundwater behave myopically (MB), they choose g_t and γ_t by individually maximizing profit, ignoring the dynamic evolution of G and Ψ.[5] The relevant first order conditions are:

$$p_y \frac{\partial y}{\partial g_t} = c(G_t), \tag{16}$$

$$p_y \frac{\partial y}{\partial \gamma_t} = p_\gamma. \tag{17}$$

Although optimal management could be expected to achieve a larger aquifer of better quality, it can be shown through comparison of (16) and (17) with (8) and (9) that the common property solution attains a steady state which might have higher or lower quantity, although it will certainly have better quality. The possibility that intervention worsens one of the objectives cannot be disregarded. This result can be explained by noting that higher quality water is more productive, so that it may be efficient to pump more of it and reach a steady state where there is less water. It can also be shown, using Equation (13), that myopic users will apply more of the polluting input than would be efficient at the steady state, which is the anticipated result.[6]

3.2. TAXES

One of the ways for the common property arrangement to replicate the optimal solution is to impose a set of input taxes on users so that their choice of both extracted water and applied contaminants reflects society's marginal costs instead of their own. In this case, that would require:

$$T_t^g = (1-\alpha)\lambda_t + \beta_t \frac{\partial e}{\partial g_t}, \tag{18}$$

$$T_t^\gamma = \beta_t \frac{\partial e}{\partial \gamma_t}. \tag{19}$$

Naturally, the tax on water reflects both its scarcity cost and its role in groundwater contamination. Moreover, although the tax on the polluting input only reflects its impact on quality (as γ does not affect G directly), it should be stressed that optimal values of β_t and $\partial e/\partial \gamma_t$ will also depend on the path of water extraction.

The tax schedule presented above would bring myopic users to optimal levels of g_t and γ_t for all t, and thus maximize social welfare. A theoretically equivalent alternative would be to issue permits for the use of water as well as for the other input. The market price of water permits would reflect both water's relative scarcity and its effect on quality once applied. Conventional management arrangements, where only quantity or only quality are controlled, are presented subsequently to emphasize their inefficiency.

Consider first the problem of the 'Stock Manager', who takes quality parameters as given and seeks to impose a tax on extracted water. The first order conditions for extraction yield:

$$p_y \frac{\partial y}{\partial g_t} = c(G_t) + \lambda_t (1 - \alpha), \tag{20}$$

$$\dot{\lambda} = \rho \lambda_t + M \frac{\partial c}{\partial G_t} g_t. \tag{21}$$

If the manager takes γ and Ψ as given and sets a tax of $\lambda_t(1 - \alpha)$, not only is this solution different from the optimal one, it may also lead the aquifer towards a steady state that is even further from its optimal state (if it was the case that $G^{MB} > G^{OPT}$). With the tax, first order conditions for the user's problem are (17) (which the regulator ignores) and (20). The imposition of this tax will alter his decisions on g_t and γ_t. However, in the steady state extraction returns to its usual exogenous value, so that Ψ and γ must also return to their previous steady state levels (by (17) marginal revenue of γ is constant at all times as long as the prices of output and γ remain constant). Thus Equation (20) implies that the new steady state stock is higher than the myopic case. The selected tax is inefficient because it does not consider water's role as a contaminating vector. However, note that this does not mean the tax is always lower than the optimal one, since λ_t as calculated by the Stock Manager is different.

Now consider the problem of the 'Contamination Regulator', who will decide what taxes to charge on contaminating inputs, considering that water is pumped at a constant price (or equivalently, that G is taken as given). The choice will be characterized by:

$$p_y \frac{\partial y}{\partial g_t} = c(G_t) + \beta \frac{\partial e}{\partial g_t}, \tag{22}$$

$$p_y \frac{\partial y}{\partial \gamma_t} = p_\gamma + \beta \frac{\partial e}{\partial \gamma_t}, \tag{23}$$

$$\dot{\beta} = (\rho + \delta) \beta - M p_y \frac{\partial y}{\partial \Psi}. \tag{24}$$

Thus chosen taxes will again not coincide with the optimal ones. In this case not much can be said on the evolution of variables after taxes are set. Nevertheless, it is clear that the optimal stationary solution will never be achieved because with similar values of γ, Ψ, g, and G, Equations (8) and (22) cannot both be valid. Moreover, if both types of regulators exist and taxes are set separately, the optimal solution will generally still not be achieved, and results will depend on what sort of strategy each of the regulators follows. This may have important implications both for policy and for institutional design.

4. Minimum Quality Requirements

Previous sections assumed that the main reason for controlling groundwater pollution was its impact on the productivity of applied water, since it was an input in the production of some other good. However, the observed necessity for pollution control has often arisen from a different reason. Water contamination brings other damages to society, either directly (public supply of drinking water) or indirectly (deterioration of ecosystems associated with aquifers). To include these effects in the previous model, it would be sufficient to come up with a water damage function for society, add it to the objective function of the model, and proceed with maximization as usual, calculating optimal taxes that would induce polluters to act efficiently.[7] This could easily be done if the water damage function was known. Since it generally is not, an alternative is to impose existing quality requirements on the management problem.[8]

In what follows it will be assumed that quality does not affect production. It will be shown that even in this case, the existence of a restriction on Ψ means that optimal policy must again consider stock effects on quantity and quality jointly. The management problem is now:

$$\max_{g_t, \gamma_t} \int_0^\infty M\left[p_y y(g_t, \gamma_t) - c(G_t) g_t - p_\gamma \gamma\right] e^{-\rho t}\, dt, \tag{25}$$

subject to the same restrictions as problem (6) plus an additional one: $\Psi_t \geq \underline{\Psi}$, where $\underline{\Psi}$ is the minimum permissible quality level. Associating a multiplier function, $\eta(t)$, with the restriction, yields:[9]

$$p_y \frac{\partial y}{\partial g_t} = c(G_t) + \lambda_t (1 - \alpha) + \beta_t \frac{\partial e}{\partial g_t}, \tag{26}$$

$$p_y \frac{\partial y}{\partial \gamma_t} = p_\gamma + \beta_t \frac{\partial e}{\partial \gamma_t}, \tag{27}$$

$$\dot{\lambda} = \rho \lambda_t + M \frac{\partial c}{\partial G_t} g_t, \tag{28}$$

$$\dot{\beta} = (\rho + \delta)\, \beta_t - \eta_t, \tag{29}$$

$$\eta_t \geq 0; \quad \eta_t (\Psi_t - \underline{\Psi}) = 0. \tag{30}$$

These look exactly the same as in the case where productivity of water depended on its quality, except the behaviour of quality's shadow price is explained by a different reason: the opportunity cost of diminishing quality no longer reflects its impact on profit, but it incorporates the effect of the restriction. Just as before, optimal policies include effects both on quality and on quantity. If the restriction is never binding (i.e. quality is never a problem), then $\eta_t = 0, \forall t$, implying that $\beta_t = 0$, so that only quantity actually needs to be managed. In this case, typical quantity-only water resource models would be sufficient.

On the other hand, it can be shown that the restriction will be binding at the steady state if $\underline{\Psi}$ is higher than Ψ^{MB}. To show this, note that if $\underline{\Psi}$ was not binding at the steady state, the stationary value of β would be zero, so that by (27) the chosen value of γ would be the myopic one. However, with the same g and γ, steady state quality would be the same as that of the myopic case, so that $\underline{\Psi}$ must in fact be binding. Thus, if an aquifer that is being exploited under the common property regime has a quality problem, better management of water quantity will not solve the quality problem.

The fact that $\underline{\Psi}$ is binding at the steady state may allow policy makers to look at the problem in a slightly different way. Since $\underline{\Psi}$ can be used to calculate the corresponding steady state stock $\underline{G}$, using Equation (26) and replacing λ with its steady state value, there is a combination $(\underline{\Psi}, \underline{G})$ which can be used as the joint management goal. Thus the cost-effectiveness of alternative instruments (especially suboptimal ones) in reaching this goal can be evaluated. The main difference for the dynamic case is that the quality restriction may not be achievable as soon as it is imposed, so that a few periods will lapse before the target is reached (the choice of actions to undertake during this period may consider criteria related to the speed of adjustment associated with alternative policies). The illustration in Section 5 includes such considerations.

Although it is theoretically easy to calculate a set of optimal taxes that solve the groundwater management problem, the implementation of such a scheme is quite a different matter. Not only might there be monitoring problems, but in reality, the M users of the aquifer will generally not be all alike, either because their production processes are different or because the pollution functions are different. Such disparities do not complicate the theoretical model much, since optimal taxes will look the same except that there are M sets of individual ones. However, implications for implementation are huge. Furthermore, if different users have different quality requirements, the dynamics of the model might imply that some users abandon groundwater extraction altogether as its quality worsens.

Helfand (1999) and Helfand and House (1995) analyze problems with input-based policies and compare them to emissions-based ones, noting that the latter are more difficult to implement in the nonpoint pollution case due to the frequent unobservability of emissions. The implementation of emissions taxes is even less attractive in the joint quantity/quality dynamic model because water has to be taxed anyway due to the scarcity problem. These two papers, as well as the paper by Larson et al. (1996) deal specifically with the issue of suboptimal input-based instruments, such as uniform taxes for different users or single-input taxation. Additional inefficiencies of simpler schemes might appear in the dynamic case, since optimal tax levels require continuous adjustment to reflect changing scarcity and quality costs.

5. Illustration

In this section an application in the management of a contaminated aquifer is developed.[10] This example was constructed using real, estimated functions from several sources, namely Yadav (1997) (contamination function), Larson et al, (1996) (production and emissions functions), as well as Zeitouni and Dinar (1997) and Feinermann and Knapp (1983) (some aquifer characteristics, pumping costs). It analyzes agricultural production using nitrates and irrigated water under myopic behaviour of farmers and compares it to a quantitatively optimal solution. Then a minimum quality requirement (or rather, a maximum concentration limit) is imposed and the optimal taxes that would lead the common property equilibrium to an efficient solution that satisfies the restriction are identified.

The notation is slightly different from that of previous sections because empirical applications use pumping lift and pollutant concentration, rather than the more theoretical variables G and Ψ. However, there is one to one correspondence between lift and stock, on one hand, and quality and concentration, on the other. Moreover, the profit maximizations will be undertaken per ha instead of per farmer.

The evolution of lift, L (which represents the distance between the land surface and the water level) can be represented by:

$$\dot{L} = \frac{(1-\alpha)g^T - R}{AS}, \tag{31}$$

where g^T is the total water used (g^T is the sum for the entire area farmed of the per ha water applied) , A is aquifer area and S is the specific yield. Pollutant concentration will be assumed to evolve according to:

$$\dot{C} = \eta \mathrm{NO}_3 - \delta C_t,$$

where NO_3 is emitted nitrates as a function of applied nitrogen and water, and η is a scaling term. The production function is quadratic and unit pumping cost is linear in lift. Tables 1, 2 and 3 present parameter values and clarify physical units.

Starting with the aquifer in a 'pristine' condition ($L(0) = 1$, $C(0) = 0$) and solving the maximization problem without considering any quality restrictions, the appropriate first order conditions (see Equations (16) and (17)) yield the myopic behaviour paths for lift, water and nitrates (Figures 1, 2 and 3). At the steady state, $\gamma = 88.953$, $g = 611.08$, and $L_t = 101.05$. Furthermore, the aquifer will be contaminated as soon as three years after initial exploration, and pollutant concentration will reach a steady state value of $C = 62.17$ (concentration worsens even more in

Table 1. Model parameters.

Variable		unit	value*
p_Y	Crop price	$	500
Y	Crop output	ton/ha	*tbd*
γ	Applied nitrogen	kg/ha	*tbd*
g	Applied water	mm	*tbd*
z	Pumping cost	$/mm-ha/m	0.003
L	Pumping lift	m	*tbd*
p_N	Price of nitrogen	$	0.4
AF	Area farmed	ha	1000
alfa	Irrigation return rate		0.1
R	Recharge	m^3	5.50E+06
A	Aquifer area	m^2	20000000
S	Specific yield		0.1
NO_3	Nitrates leached	kg/ha	*tbd*
C	Pollutant concentration	mg/l	*tbd*
heta	Scaling factor		0.13882632
delta	Decay rate		0.2
ro	Discount rate		0.03

**tbd* means that the variable is determined by the model.

Table 2. Production function.

$Y = a + bN + cW + dNW + fN^2 + gW^2$	
a	2.52
b	0.000535
c	0.00151
d	0.000002
e	–5.38E-06
f	–8.85E-07

Table 3. Leaching function.

$NO_3 = h + IN + jW + kNW$	
h	–26.06
I	–0.152
j	0.158
k	0.0006

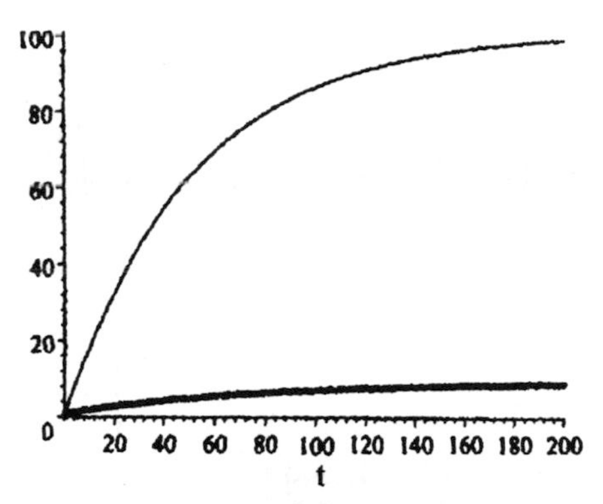

Figure 1. Pumping lift (m).

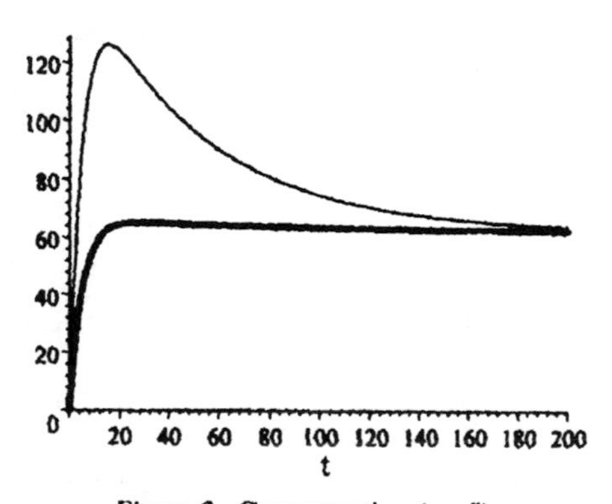

Figure 2. Concentration (mg/l).

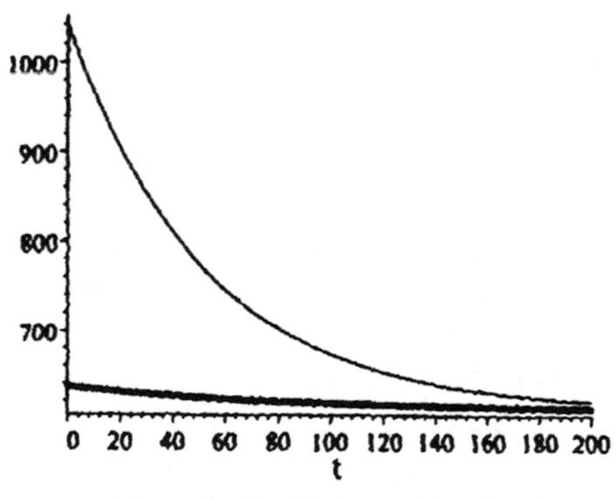

Figure 3. Applied water (mm).

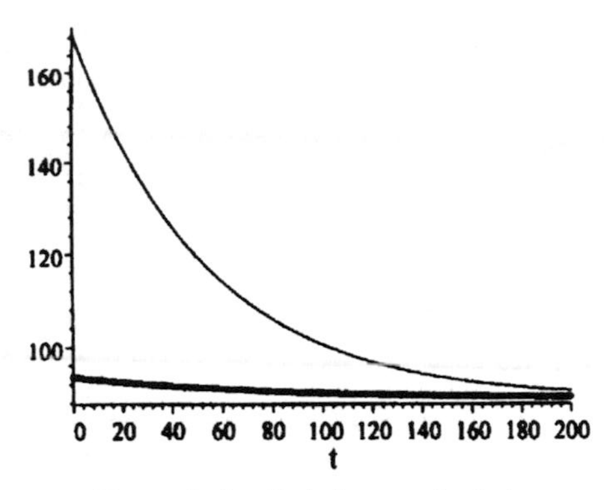

Figure 4. Applied nitrogen (kg/ha).

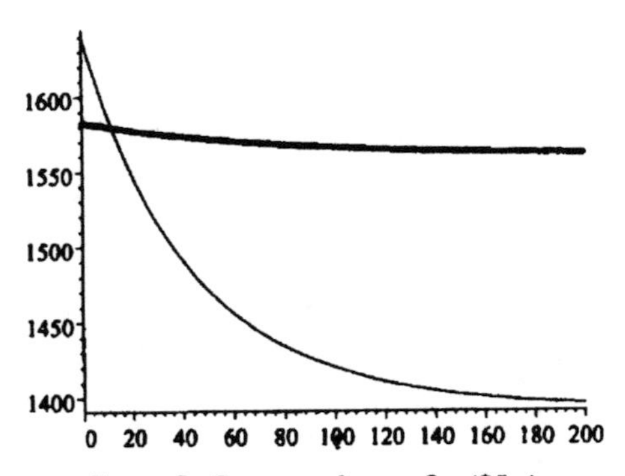

Figure 5. Current value profits ($/ha).

the initial periods, but as the aquifer table becomes lower, pumping and fertilizer applications decrease until C is stable, see Figure 2).

In contrast, optimal quantitative management paths are derived from Equations (8) and (9).[11] At the steady state, $L = 9.382$, $g = 611.1$ and $\gamma = 88.955$. The steady state is achieved asymptotically and the present value of total welfare losses from common property management amounts to a negligible $1461/ha (which is approximately 2.8% of optimal welfare). The reason is that lift is small in the first years of pumping, so costs are similar; yet more nitrogen and water are applied in the common property case, making revenue higher (see Figure 5). Because of discounting, these first periods weigh heavily on the final outcome. Only after some periods of aquifer overexploitation does lift decrease sufficiently to make this effect disappear, thereby yielding lower profits for common property than those of the efficient case. The comparative evolution of relevant variables is in Figures 1–5 (optimal paths are the thick ones).

This looks like one of those aquifers for which quantity only management does not bring great welfare gains, in the line of Gisser and Sanchez (1980) and others. However, considering current restrictions for maximum allowable concentration

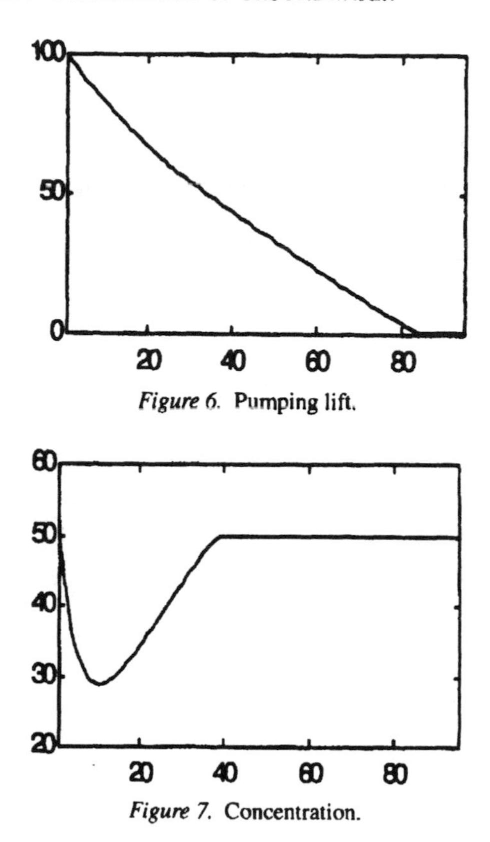

Figure 6. Pumping lift.

Figure 7. Concentration.

of nitrates in the water of 50 mg/l, this aquifer clearly has a quality problem. How might that problem be solved in our example? The actual answer depends on when the restriction is imposed. It could be imposed as soon as concentration hits 50 mg/l, which will happen between year 3 and 4, but this would imply the aquifer would never be overpolluted. More realistically, it can be assumed that agents were already overpolluting the aquifer before some measure is attempted. For ease of exposition, take as a starting point the common property steady state described above. If nitrogen applications are forbidden until the standard is reached (which has a much lower cost than forbidding cultivation altogether), it will take between three and four years to do so. Farmers could be compensated for their profit losses, which would amount to 52.8 $/ha for the whole period. At that point, $L = 99.477$ and $C = 50$. After that time, a set of quantity-quality taxes would be imposed both on water and fertilizer. The relevant system of differential equations cannot be solved analytically, and it does not have an interior steady state. A solution was obtained using MATLAB.

Optimal paths for lift, water, fertilizer, and concentration are presented in Figures 6–9. Concentration actually falls below its maximum allowable value for the

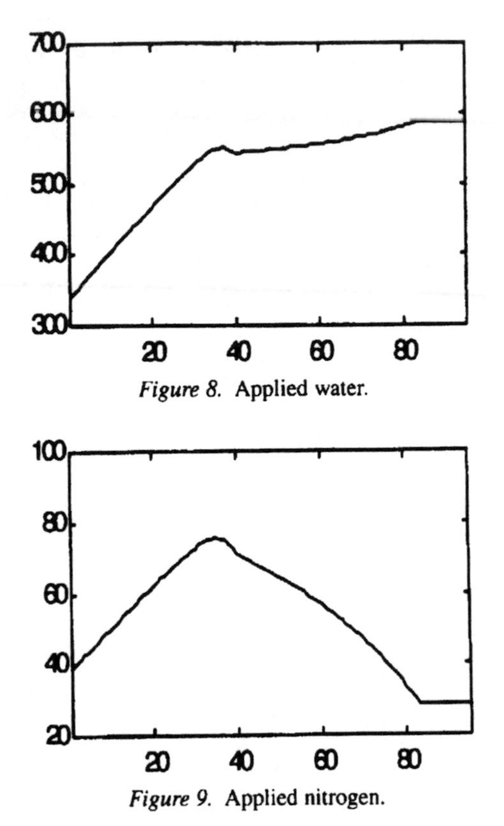

Figure 8. Applied water.

Figure 9. Applied nitrogen.

first few periods, where the incorporation of the 'quantitative' opportunity cost leads to much lower water use than that of the MB steady state. Then, as lift decreases, pumping and fertilizer increase, until the concentration limit is reached again. The average tax on water is \$0.23/mm-ha and that on fertilizer is \$0.11/kg (although it is zero as long as $C < 50$ mg/l) (see Figures 10 and 11). The difference in steady state welfare between the optimal case with and without the quality restriction is a paltry \$1/ha per year. That is the cost the government is imposing on society by requiring a minimum quality level.

6. Conclusion

Economic literature on groundwater management has traditionally been split into two areas: on the one hand there are papers that evaluate different schemes of dynamic aquifer management, considering that pumping costs vary with stock but ignoring water quality. On the other hand there are papers that consider contamination problems caused by specific pollutants. Among these, only a few are dynamic

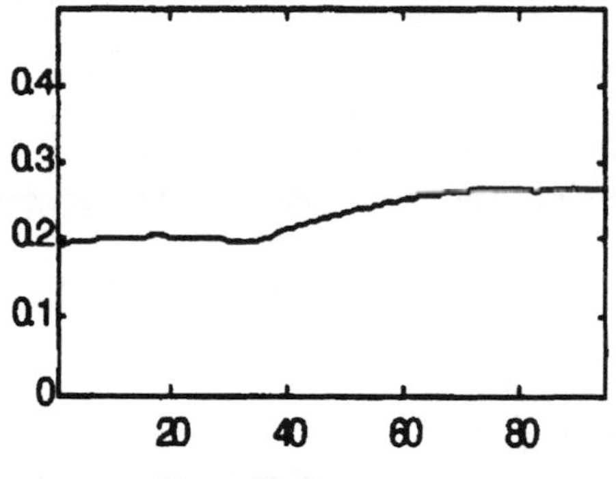

Figure 10. Tax on water.

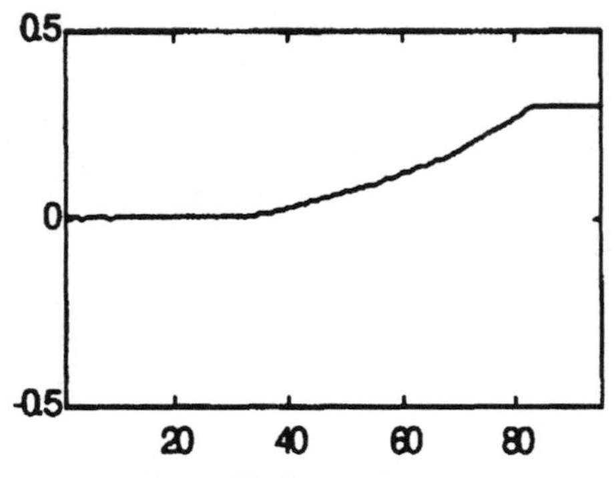

Figure 11. Tax on fertilizer.

in nature, and only a few include water in their estimation of pollution functions. Yet even those consider that water can always be obtained at a constant price, which does not reflect changes in pumping costs. This chapter presents two alternative models for joint quality-quantity management, and it shows that existing models are in fact special cases of these. Thus existing models continue to be adequate when aquifers conform to those special cases, but are in general not applicable to the more complex problem of managing an aquifer when both quantity and quality are relevant variables.

Using assumptions from both areas of the literature, the chapter shows that when quality and quantity of aquifers are important, optimal policies must reflect the relationship between them. The main features used in the characterization of production and pollution functions draw heavily on agricultural contamination literature. Furthermore, since the myopic behaviour hypothesis is adopted, the model is particularly suitable to analyze aquifers exploited by a large number of agricultural producers that are relatively small and face identical physical conditions. Nonetheless, the same framework could be adapted to other types of common property behaviour as well as to different uses for the water.

The theoretical model furnishes some interesting insights, but those focus mainly on steady state characteristics. Specific functions for production, pollution production, and pumping costs are required for a more complete analysis of system properties. Ultimately, the usefulness of models such as these hinges on empirical work, not only to check which variable paths lead to which steady states, but also to evaluate welfare losses from different, nonoptimal situations, so that unregulated common property or regulated common property with suboptimal instruments can be appraised. That is the purpose of the illustration presented.

Other important areas have been left out in the analysis. Paramount among these are the choice between surface water and groundwater in conjunctive use models with quality differences, and the incorporation of uncertainty. When quality is an issue surface water and groundwater can no longer be considered perfect substitutes in production. Moreover, available surface water is not constant throughout time (in semi-arid climates great variability is observed both within the year and between different years). The inherent uncertainty in hydrological variables could have important implications. In fact, there is inherent uncertainty both in water availability (especially in semi-arid climates) and in many aquifer pollution processes. Further research should be dedicated to these aspects.

Notes

[1] Likewise, for surface water management quantity and quality should also be considered jointly (see Costa, 1992). However, the static framework adopted for surface water is inadequate for groundwater.

[2] Examples of papers where water quality affects productivity are Letey and Dinar (1986), Zeitouni and Dinar (1997) and Dinar (1994), although in those cases the relevant quality parameter is salinity which is not dependent on specific inputs.

[3] This specification is equivalent to $\Psi = -C$. The additional term simply ensures that Ψ is a positive value. An alternative specification is $\Psi = 1/C$.

[4] These are necessary and sufficient if the Hamiltonian function is jointly concave in g, γ, G and Ψ or if the maximized Hamiltonian is concave in G, Ψ. At this point concavity requirements are assumed to be satisfied. However, in typical empirical applications such conditions may not hold (see Section 5).

[5] For other types of common property equilibrium, see Provencher and Burt (1993), Xepapadeas (1996), Rubio (2000), and the references therein.

[6] Different results are obtained in Roseta-Palma (2002) with different assumptions on the quality function, namely that $\partial f/\partial\Psi > 0$. This disparity highlights the importance of accurately representing the physical properties of the aquifer and its reactions with specific contaminants.

[7] See Yadav (1997). To adequately choose the optimal quality level would require information on the damages brought about by increased pollution concentration, in terms of health, environmental or amenity effects.

[8] Another alternative would be to consider a cost function associated with treating water of quality Ψ whenever $\Psi \leq \underline{\Psi}$.

[9] Additionally, eventual discontinuities of the co-state variables would have to be analyzed. See for example Seierstad and Sydsaeter (1987).

[10] Many grateful thanks to Ana Balcão Reis, Duarte Brito, and especially António Antunes; without their input this illustration would have been impossible.

[11] Due to the linearity of pumping costs these are not sufficient. However, the system of differential equations defined by the first order conditions and the transition equation for L is linear, so it can be solved and there is only one stable path that satisfies first order conditions.

References

Anderson, G., Opaluch, J. and Sullivan, W.M. (1985) Nonpoint agricultural pollution: Pesticide contamination of groundwater supplies, *American Journal of Agricultural Economics* **67**, 1238–1246.

Costa, L. (1999) The interdependence and control of water quantity and quality externalities, unpublished manuscript.

Dinar, A. (1994) Impact of energy cost and water resource availability on agriculture and groundwater quality in California, *Resource and Energy Economics* **16**, 47–66.

Feinermann, E. and Knapp, K.C. (1983) Benefits from groundwater management: Magnitude, sensitivity, and distribution, *American Journal of Agricultural Economics* **65**, 703–710.

Gisser, M. and Sanchez, D.A. (1980) Competition versus optimal control in groundwater pumping, *Water Resources Research* **16**, 638–642.

Griffin, R.C. and Bromley, D.W. (1982) Agricultural runoff as a nonpoint externality: A theoretical development, *American Journal of Agricultural Economics* **64**, 547–552.

Helfand, G. (1999) Controlling inputs to control pollution: When will it work?, *AERE Newsletter* **19**(2), 13–17.

Helfand, G. and House, B. (1995) Regulating nonpoint source pollution under heterogeneous conditions, *American Journal of Agricultural Economics* **77**, 1024–1032.

Larson, D., Helfand, G. and House, B. (1996) Second-best tax policies to reduce nonpoint source pollution, *American Journal of Agricultural Economics* **78**, 1108–1117.

Letey, J. and Dinar, A. (1986) Simulated crop-water production functions for several crops when irrigated with saline waters, *Hilgardia* **54**(1), 1–32.

Mapp, H.P., Bernardo, D.J., Sabbagh, G.J., Geleta, S. and Watkins, K.B. (1994) Economic and environmental impacts of limiting nitrogen use to protect water quality: A stochastic regional analysis, *American Journal of Agricultural Economics* **76**, 889–903.

Neher, P. (1990) *Natural Resource Economics – Conservation and Exploitation*, Cambridge University Press, Cambridge.

Provencher, B. (1993) A private property rights regime to replenish a groundwater aquifer, *Land Economics* **69**(4), 325–340.

Provencher, B. (1995) Issues in the conjunctive use of surface water and groundwater, in D. Bromley (ed.), *The Handbook of Environmental Economics*, Blackwell, Oxford, pp. 503–528.

Provencher, B. and Burt, O. (1993) The externalities associated with the common property exploitation of groundwater, *Journal of Environmental Economics and Management* **24**, 139–158.

Roseta-Palma, C. (2002) Groundwater management when water quality is endogenous, *Journal of Environmental Economics and Management*, forthcoming.

Rubio, S. (2000) Strategic behavior and efficiency in the common property extraction of groundwater, Paper presented at the Symposium on Water Management – Efficiency, Equity and Policy, at the University of Cyprus, Nicosia, September 22–24.

Segerson, K. (1988) Uncertainty and incentives for nonpoint pollution control, *Journal of Environmental Economics and Management* **15**, 87–98.

Seierstad, A. and Sydsaeter, K. (1987) *Optimal Control Theory with Economic Applications*, North-Holland, Amsterdam.

Shortle, J. and Dunn, J. (1986) The relative efficiency of agricultural source water pollution control policies, *American Journal of Agricultural Economics* **68**, 668–677.

Vickner, S., Hoag, D., Frasier, W.M. and Ascough II, J. (1998) A dynamic economic analysis of nitrate leaching in corn production under nonuniform irrigation conditions, *American Journal of Agricultural Economics* **80**, 397–408.

Xepapadeas, A.P. (1992) Environmental policy design and dynamic nonpoint-source pollution, *Journal of Environmental Economics and Management* **23**, 22–39.

Xepapadeas, A.P. (1996) Managing common-access resources under production externalities, in A. Xepapadeas (ed.), *Economic Policy for the Environment and Natural Resources*, Edward Elgar, pp. 137–157.

Yadav, S. (1997) Dynamic optimization of nitrogen use when groundwater contamination is internalized at the standard in the long run, *American Journal of Agricultural Economics* **79**, 931–945.

Zeitouni, N. and Dinar, A. (1997) Mitigating negative water quality and quality externalities by joint management of adjacent aquifers, *Environmental and Resource Economics* **9**, 1–20.

Missing Markets and Redundant Reservoirs: Dams as a Consequence of Inefficient Groundwater Management Policies

BEN GROOM and TIM SWANSON

1. Introduction

The blocking of natural waterways has been argued to be one of the clearest cases of human-generated damages to natural systems and biodiversity. (World Commission on Dams, 2000). Is this damage potentially avoidable? And why are the world's waterways flooded with dams? Of course, many of these structures are erected in the pursuit of clear development benefits (energy, irrigation) and represent examples of the trade-offs existing between the pursuit of developmental and environmental benefits (Krutilla and Fisher, 1967). Others however may be the result of inefficient water management practices, and hence entirely avoidable. It is this class of wasteful construction/obstruction projects that we wish to describe and to define here. It is our argument that dams may be one consequence of inefficient groundwater management practices – a response to the persistence of externalities between groundwater owners and conjunctive users.

The existence of externalities regarding groundwater has been well documented theoretically and empirically (see, e.g., Dasgupta, 1982; Provencher, 1994; Gisser and Sanchez, 1980; Provencher and Burt, 1993). In the taxonomy of Provencher and Burt (1993) common property externalities in groundwater can be divided into three distinct varieties: *Stock*, *Depth* and *Risk* externalities. Stock and depth externalities arise respectively from the over pumping of a finite groundwater stock and the associated increased pumping depth as compared to the social optimum. Groundwater also has a value as an insurance policy against uncertain surface water supplies and hence income: the so-called *buffer value* (Tsur, 1990; Tsur and Graham-Tomasi, 1991; Provencher, 1994; Cummings, 1969). Under commonality a *risk externality* arises, when groundwater abstraction results in dissipation of stocks that might be used as buffers.

This chapter is concerned with groundwater externalities that arise *even if the property rights to groundwater are well defined but when markets are missing.* This occurs in this case if the entity with full and exclusive property rights over groundwater makes the decision on the timing of groundwater use without giving

P. Pashardes et al. (eds.),
Current Issues in the Economics of Water Resource Management, 171–192.

weight to the preferences of other users of conjoint surface water flows, due to the absence of markets spanning the property right holder and linked water users. Since groundwater usage will often translate into some impacts on surface water flows (if only because the groundwater user then uses less surface water), this implies the potential existence of externalities even in the context of well-defined property rights in groundwater supplies. The description of this case, and the outlining of its consequences, is the object of this chapter.

We examine the problem of groundwater management externalities in a stochastic and spatially-separated two-sector two-period model. In this model the two sectors are linked by a surface flow and common climate, but only one of the two sectors has rights over groundwater. We also assume that there are no water markets linking the two sectors. Given these assumptions, we then demonstrate that these are necessary but not sufficient conditions for socially sub-optimal intertemporal allocations of water. The nature of this externality is discussed, and its dependence upon the heterogeneity of preferences between the sectors is shown. The implications of such an externality are interesting. We find that when this form of risk externality exists and water markets do not, the second-best response by water users may manifest itself in potentially sub-optimal investments in insurance against the stochastic surface water. Finally, we argue that pursuing more efficient water management policies would potentially displace the need for such reservoirs – they would be redundant.

The chapter proceeds as follows. Section 2 introduces the constituent ingredients of the model while Section 3 discusses the necessary and sufficient conditions for the existence of a time profile externality. Section 4 provides a simulation of its potential impact. In Section 5 it is shown that the existence of time profile externalities provides incentives for sub optimal investment in smoothing capital, e.g. a dam.

In conclusion, this chapter highlights the problem of time of use externalities, where interdependent water users care about the timing of water use as well as the amount. It describes an unusual circumstance in which well-defined property rights in groundwater may nevertheless result in suboptimal water management when conjoined surface water users are also considered (and markets between them are missing). Finally we argue that the socially optimal management of groundwater may postpone or avert entirely costly investments in large capital outlays such as storage dams. Optimal management of groundwater may realise the buffer values of groundwater resources rather than having those values capitalised in investments in smoothing capital. This would allow the world's waterways to avoid a few of the least needed dams.

2. The Model: Stochastic Surface Water, Spatial Separation and Heterogeneous Risk Preferences

We use a two period model with two differentiated sectors to examine the case of property right failures when water markets are incomplete. We wish to focus solely on the problem of conflicts over preferences for 'time of use profiles', and when externalities in time of use will exist. The components of the model are: (a) stochastic surface flows; (b) heterogeneous preferences over the resulting uncertainty; and (c) spatial separation.

2.1. The Stochastic Surface Water Model

Following Provencher and Burt (1993), it is assumed that there are two periods, designated 0 and 1. The aquifer has a stock x and groundwater abstraction per firm is u at a cost of $c(x)$. There is a stochastic delivery of surface water in period 1 of q. Available groundwater in the terminal period is defined by the state equation:

$$Z = x_0 - U_0. \tag{1}$$

Z is essentially the 'carryover' of the groundwater stock in the terminal period from period 0, and U_0 is the total groundwater usage in period 0, by the M firms in the system. Net firm income in the period 0 and period 1 is described respectively by:

$$h(u_0) - c(x_0)u_0, \tag{2}$$

$$h(u_1(Z, q), q) - c(Z)u_1(Z, q). \tag{3}$$

This indicates that the income in the terminal period is stochastic due to the dependence on the stochastic variable q, and conditional on Z.[1,2]

2.2. Heterogeneous Risk Preferences

Given the stochastic nature of the surface flow in the terminal period, water users will experience uncertainty in regard to their future incomes. It is assumed that users reduce their valuation of expected income flows in the terminal period by an amount which reflects their individual risk perceptions, and that such risk characteristics may be captured in individual risk premia. Hence, expected utility of an uncertain surface flow is expressed, where π is net present value of income, as follows:

$$E\{v_0^i(\pi)\} = E\pi - \rho_i, \tag{4}$$

where ρ_i represents the user's *risk premium*. The risk premium of the user will be a function of the variance, and perhaps higher order moments of the terminal income and characteristic *risk aversion*. Each user applies a distinct risk premium to groundwater as a result of this aversion to income risk and the capacity of groundwater to smooth supplies over time. Clearly the size of the premium will depend upon the preferences for secure water in the terminal period.

The risk premium is assumed to be a function of wealth, represented here by the initial period's income, and other assets available for smoothing uncertain flows, including the terminal period groundwater stock, Z. Therefore the risk premium of the individual i, is described by the function:

$$\rho_i = \rho_i(u_0, Z). \tag{5}$$

The importance of heterogeneity in society's risk preferences is that each individual risk premium will contribute to the determination of an individual's preferred *time profile of groundwater use*. That is, given differing preferences, different individuals would possess different time profiles for the allocation of groundwater: *ceteris paribus* a more risk averse individual will prefer to allocate more groundwater toward the uncertain terminal period than would the less risk averse individual.[3]

2.3. Spatial Separation: Conjoint Use and Missing Markets

The final element of the model requires two spatially separated water users (or 'sectors'). This spatial separation has two dimensions, one hydrological and one legal. First, the spatial separation allows for the users to be conjoint in the use of the surface flow and the groundwater resource. This mutual interdependence may be very simple in nature, e.g. one sector's reduced use of surface flows by reason of increased extraction of groundwater, or it might be more complicated (e.g. by reason of return flows). All that is required for our purposes is that one sector's decision about the time profile of groundwater extraction impacts on the other sector's time profile of water availability.

The legal element of the model concerns the property right structure regarding groundwater resources and the resulting control over the choice of the prevailing time of use profile. Most jurisdictions invest the owner of the overlying lands with the right to determine the time profile of groundwater extraction, irrespective of its impacts on the availability of water to downstream users.[4] Hence spatial separation also implies that control over groundwater extraction may lie with one sector to the exclusion of another.

It is also important to assume that the legal control over groundwater resources is not transferable, i.e. that there is no spatial or futures market in groundwater. This may be attributable to legal restrictions or to market imperfections. In short, we will be looking at the problem initially from the second-best perspective, in which there are missing markets between water users impacted by conjunctive groundwater use.

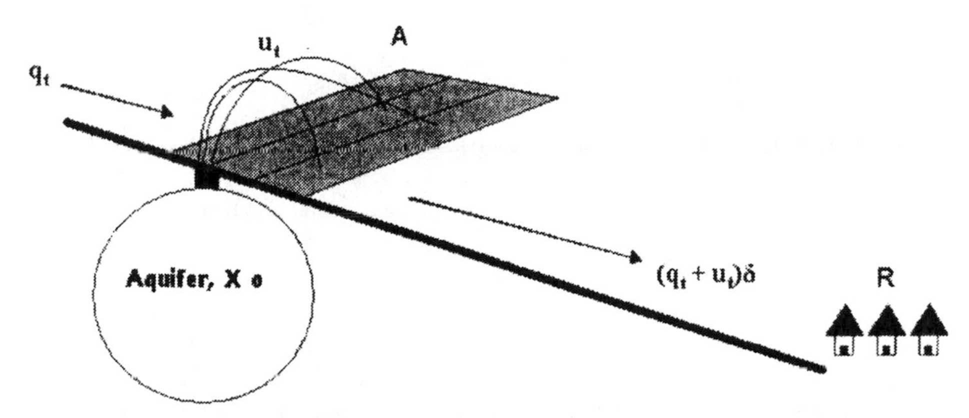

Figure 1. The spatial characterisation of this conjunctive use model.

2.4. A Conjunctive Use Model

We will now specify a particular form of two-sector spatial model, incorporating the types of hydrological interaction and legal institutions described above. The defining characteristic of the model is spatial: the upstream sector is deemed to lie above the groundwater resource and wholly upstream from the second sector. This definition implies two important features. First, given that the property rights to the resource are based on location, the upstream sector has control over the quantity and timing of the pumping of groundwater. Secondly, as the 'downstream user', the second sector's access to the surface water is dependent upon the pumping behaviour of the upstream sector. This is of course due to the fact that both users have access to the surface water, but the upstream user has access to both resources.

The nature of this interaction may take many forms, but we will select a simplified form of interaction that allows us to focus on the time profile of use externality. We will assume that when the upstream sector pumps groundwater and makes use of it, it is converted into a surface flow and flows ultimately to the downstream sector by way of an ephemeral river. This spatial inter-dependence is characterised in Figure 1. The factor δ denotes the spatial externality involved in the decision making by the sector in control of the aquifer. When the upstream sector, A, pumps an amount of water u from the aquifer, a surface flow of δu_t is generated for use by the downstream sector, R (where $0 \leq \delta \leq 1$). Thus the upstream user's time of access to groundwater determines – in part – the timing of the availability of surface water to the downstream user. In order to focus on the conflict of resource use that arises as a result of the timing of the resource availability, rather than conflicts over exclusive use of the water, it will be assumed that $\delta = 1$.[5]

2.5. The Kouris Catchment Area in Cyprus: An Example

To render our discussion more concrete, it may be instructive to use a specific case study region to illustrate the particularities of our model. The Kouris catchment area in Cyprus commences in the high precipitation Troodos Mountains of central Cyprus. Water flows from the higher plateau region down toward the lowlands and on to the Mediterranean coastline. In the uplands at the foot of this mountain range there is a substantial and varied agricultural industry, including vineyards, fruits and vegetables. It is also in this region that the aquifers in the catchment area exist. Nearly all of the aquifers lie beneath and are controlled by the agricultural sector. Further down this watershed, coastal tourism and urban development have dramatically increased the population along the coastal reaches of the Kouris basin. In these areas there are no aquifers, and so the burgeoning urban population is entirely dependent upon surface water flows.

Recent research has shown a strong dependence between the surface water flows that reach the coast and the groundwater controlled largely by agriculture; approximately 60% of surface water flows have been found to be sourced in groundwater supplies further upstream (Boronina et al., 2001). Extreme variability in water supply implies that this source could be retained for water smoothing and as a buffer, but this is currently not the case. During the wet winter season, significant amounts of surface water flow exists, but during the dry summer season little or no surface water reaches the coastal residential population in this catchment area. In order to store surface water for the dry season, the residential sector has invested in a substantial reservoir project that stores wet season runoff for the dry season.

In order to make pedagogical use of the Kouris catchment case study, henceforth we will refer to the upstream sector as the agricultural sector and the downstream sector as the residential sector, but do so only to emphasise that the two sectors are distinct both spatially and in other important characteristics.

3. Defining Time Profile Externalities

In this section, two scenarios are compared. In the first scenario the management of the aquifer in the face of stochastic water supply is assumed to be under optimal central control. This result is then compared to a decentralised scenario in which it is assumed that the agricultural sector acts alone, optimising the use of the groundwater resource over the two periods to satisfy its own objectives, without consideration of the downstream users. These two scenarios are contrasted and comparisons drawn between the outcomes in each case under different assumptions regarding the relative importance of the objectives of the sectors themselves and their respective risk premia.

3.1. Centralised Control of the Conjoined Surface and Groundwater

In the centralised problem the social value function is an aggregation of the values from water available and used by the agricultural sector and the residential sector. The *upstream agricultural sector* (a) overlies the aquifer and abstracts the groundwater, which is then available for non-exclusive use by both the upstream sector and the *downstream residential sector* (r). Each sector can be thought of as a single entity, each of which is distinct in its use values and risk characteristics. Subsequently the flow of water from upstream to downstream can be formally characterised as above, $u_t^a = u_t^r$ ($t = 0, 1$; a = agriculture, r = residential).[6] The social valuation of the objectives in each sector is reflected by the weights α and $(1-\alpha)$ for the agricultural sector and the residential sector respectively. The central control problem can be written as follows:

$$\begin{aligned} L &= \max_{u_0}[\alpha[h^a(u_0^a) - c(x_0)u_0^a - \rho^a(u_0^a, Z)] + (1-\alpha)[h^r(u_0^r) - \rho^r(u_0^r, Z)]] \\ &\quad + \beta E\{\alpha[h^a(u_1^a(Z)) - c(Z)u_1^a(Z)] + (1-\alpha)\cdot h^r(u_1^r(Z))\} \\ &\quad + \lambda_0(x_0 - u_0^a). \end{aligned} \tag{6}$$

The solution of this problem is characterised by Equation (7) and must satisfy the Kuhn–Tucker conditions described in Appendix A:

$$\begin{aligned} u_0^*:\ & \alpha\left[\frac{\partial h^a}{\partial u_0^a} - c(x_0) - \frac{\partial \rho^a}{\partial u_0^a}\right] + (1-\alpha)\left[\frac{\partial h^r}{\partial u_0^a} - \frac{\partial \rho^r}{\partial u_0^a}\right] \\ & = C_S^* + C_D^* + C_R^*, \end{aligned} \tag{7}$$

where

$$\begin{aligned} C_S^* &= \beta E\left\{\left[\alpha\left[\frac{\partial h^a}{\partial u_1^a} - c(Z)\right] + (1-\alpha)\cdot\frac{\partial h^r}{\partial u_1^a}\right]\frac{\partial u_1^a}{\partial Z}\right\}, \\ C_D^* &= -\beta E\left\{\alpha\left[\frac{\partial c(Z)}{\partial Z}\cdot u_1^a(Z)\right]\right\}, \\ C_R^* &= -\left(\alpha\cdot\frac{\partial \rho^a}{\partial Z} + (1-\alpha)\cdot\frac{\partial \rho^r}{\partial Z}\right). \end{aligned} \tag{8}$$

Following Provencher and Burt (1993) the marginal user cost contains three components; the stock cost C_S^*, the depth cost C_D^*, and the risk cost C_D^* (where $*$ represents the optimal solution).[7] The stock cost reflects the cost of foregone future income possibilities arising from the reduction of available water in the period 1 due to abstraction of groundwater in period 0. The depth cost describes the marginal increase of pumping costs due to abstraction in period 0. The risk cost reflects the marginal increase in the riskiness of income in period 1 that arises from marginal abstractions in period 0, and varies inversely with the level of available groundwater in period 1, Z.

Equation (7) describes the manner in which the central planner will allocate groundwater both between sectors and between time periods in order to maximise social welfare. In brief it indicates that each unit of groundwater would be allocated to the highest valued use, as between current and future uses, and as between the two sectors – according to the societal weights placed upon their respective objectives. Furthermore, in a region with distinct wet and dry seasons, and demand for water that is both constant across these seasons and averse to fluctuations, the centralised solution would involve the allocation of groundwater toward the dry season and the more risk averse sector.

3.2. Decentralised Solution – Sectoral Control over Buffer Stock

This section assumes a decentralised solution to the groundwater allocation problem described in Equation (6). In this scenario, the upstream sector controls the system through its choice of the optimal groundwater allocation from its own perspective, i.e. u_0^{a*}. In order to focus on the time profile externality, it is further assumed that property rights are well defined such that the agricultural sector behaves as a single irrigation district in its control of groundwater abstraction. In this case the decentralised solution of the groundwater allocation solution may be characterised by reconsidering (7) assuming that $\alpha = 1$:[8]

$$u_0^{a*} : \left[\frac{\partial h^a}{\partial u_0^a} - c(x_0) - \frac{\partial \rho^a}{\partial u_0^a}\right] = C_S^a + C_D^a + C_R^a, \tag{9}$$

where

$$C_S^1 = \beta E\left\{\left[\frac{\partial h^a}{\partial u_1^a} - c(Z)\right]\frac{\partial u_1^a}{\partial Z}\right\},$$

$$C_S^2 = -\beta E\left\{\left[\frac{\partial c(Z)}{\partial Z} \cdot u_1^a(Z)\right]\right\},$$

$$C_S^3 = -\frac{\partial \rho^a}{\partial Z}.$$

That is, in this case the three cost terms on the right hand side (C_S^a, C_D^a and C_R^a) represent the stock, depth and risk components of user cost *to the agricultural sector*. Cost considerations of groundwater allocations concerning other conjunctive sectors are external to the analysis.

3.3. Time Profile Externality

The difference between the two control scenarios simply reduces to the weighting that is given to the objectives of the respective sectors. In the centralised problem the solution is a weighted average of the two sectors objectives, while in the decentralised the first sector's objectives *dictate* the solution. In what follows we discuss the extent to which this difference matters.

It is not always the case that a dictated solution is socially suboptimal. If the relevant characteristics of the two sectors are identical in every respect, then it is clear that the time profile of use selected by the upstream sector will be identical to that preferred by the downstream sector. In such circumstances the decentralised management outcome will not diverge from the social optimum, the timing of the flow of water will mirror the optimum, and the identity of the controller is not important. In terms of the model set forth above, if the risk premiums and the net benefit functions in each sector are equal, the solutions to both centralised and decentralised problems are the equivalent, i.e. the solution to (7) will be the same as the solution to (9).

The corollary of this finding is that, even though property rights are well defined in the decentralised scenario, the resulting groundwater allocation will differ in general from the centralised allocation if upstream and downstream sectors' preferences differ in regard to the preferred inter-temporal allocations of groundwater.

DEFINITION (Time Profile Externality). We define a time profile externality as any situation in which one agent determines the time profile of use of a resource for another agent, and this determination is suboptimal. One agent determines another's time of use profile if they are conjunctively linked, but legally disjoint in the sense that one agent controls the time of access for both. Such a choice is suboptimal only if the agents are sufficiently distinct in relevant characteristics so that their preferred time profiles of use are distinct, and if society values all of these preferences. Such an externality persists only if the markets for its internalisation are missing.

What is the impact of a time profile externality? In this model, the groundwater resource is one possible instrument for the management of the risk imposed by stochastic surface water flows. In our construction it is possible for groundwater allocation to perform this role for the conjunctive sector at no additional cost to the controlling sector (because $\delta = 1$). However, if the two sectors are sufficiently distinct in their preferences with regard to time profile of groundwater use and markets are missing, the controlling sector may not take into consideration this function of groundwater when determining its allocation.

RESULT (The Impact of Time Profile Externality: Sub-Optimal Insurance). Other things remaining equal: if the agricultural sector is more risk averse than the residential sector, it saves more groundwater for the dry period 1 than would be preferred by the residential sector. The opposite is true if the residential sector is more risk averse than the agricultural sector. Respectively these scenarios lead to over and under-insurance compared to societal preferences.

Condition for under-insurance:

$$u_0^{a*} > u_0^* \quad \text{if} \quad \rho^a(u_0^a, Z) > \rho^r(u_0^r, Z). \tag{10a}$$

Condition for over-insurance:

$$u_0^{a*} < u_0^* \quad \text{if} \quad \rho^a(u_0^a, Z) < \rho^r(u_0^a, Z). \tag{10b}$$

Proof. Compare (9) and (11). □

3.4. Time Profile Externalities and Missing Markets – Implications for Property Rights

A time profile externality is distinct from some other forms of externality in that, even if property rights are well defined, resource use may yet be socially sub-optimal. This is because property rights require complete and well-defined markets for their efficient allocation. When the good is diverse in characteristics that are difficult to trade, such as the time of use, then even well-defined property rights will be unable to achieve efficient allocations. The absence of complete and well-defined markets in all facets of the resource implies that such social inefficiencies will exist.

In the context of water, there will necessarily be many facets of the resource that are unaccounted for in markets, because the interdependencies between users of the resource may be many and complex. The hydrological cycle links the many users together in respect to current and future uses, in ways that most of the users are probably unable to comprehend. The creation of markets for all of these many facets of this resource is probably infeasible.

In addition, the above discussion illustrates the importance of ensuring that, where control of a resource is spatially separated from certain beneficiaries, the property rights to control of the resource are most effectively vested in institutions or sectors which adequately reflect societal preferences. Sectors which adequately represent society are most likely to achieve societal goals. This is akin to a demand-side perspective on 'efficient property rights theory' (Hart and Moore, 1990). If rights are costly to relocate, then the site of their initial placement becomes more important.

Table 1. The assumed functional forms and parameterisation.

	Function/Parameter	Value
Utility function for sector i	$\frac{u_i^{(1-\theta_i)}-1}{1-\theta_i}$	-
Agricultural risk parameter	θ^a	0.2
Residential risk parameter	θ^r	0.8
Marginal cost groundwater pumping	constant	0.1
Discount rate	β	0.95
Social weight for agriculture	α	0.5
Social weight for residential	$(1-\alpha)$	0.5
Surface water in period 0	q_0	100
Expected surface water in period 1	q_1	50

4. Simulation of the Spatially Differentiated Groundwater Management Model

In order to demonstrate the mechanics of the time-profile externality and the differences between the allocations of groundwater in the decentralised and centralised cases a simple simulation is undertaken. In order to highlight the dependence of the externality on the heterogeneity of preferences for risk each sector is modelled with a different degree of risk aversion: the residential sector is modelled as being more risk averse than the agricultural sector.

4.1. Parameterisation of the Model

The simulation uses Constant Relative Risk Aversion (CRRA) utility functions for both the agricultural and residential sectors thereby explicitly determining the preferences for risk rather than expressing the expected utilities in the more general way described in (4).

Figure 2 shows the inter-temporal allocation of groundwater over a range of groundwater endowments under three different management regimes given a surface water allocation of 100 units in period 0 and an expected value of 50 units in the dry period 1. The Social Optimum represents the centrally controlled management of the groundwater resource described in Section 3.1. The Residential and Agriculture scenarios describe two decentralised management scenarios in which each of these sectors has exclusive control over time of use.

Figure 2 shows that acting individually each of these sectors would allocate units of water between periods in different ways according to their specific preferences. The residential sector is more risk averse than agriculture and has a greater preference for saving groundwater for use in the future risky time period. Con-

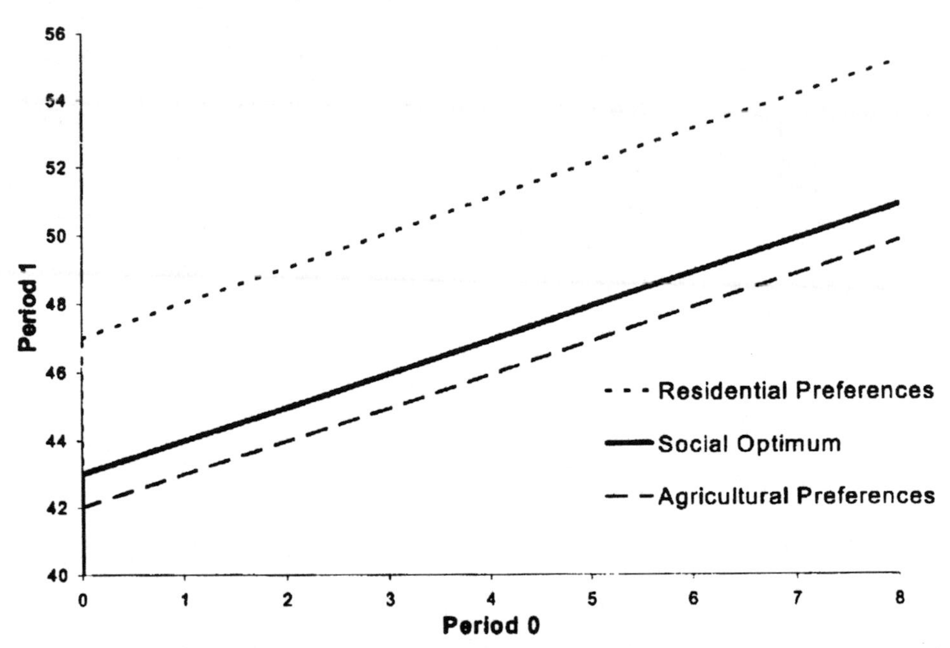

Figure 2. The inter-temporal allocation of groundwater under different management regimes and groundwater endowments.

versely the agricultural sector is less risk averse and would prefer to use more of the groundwater endowment in period 0. The Social Optimum lies between the two decentralised allocations demonstrating that it is indeed some *weighted average* of the two decentralised scenarios.

The disparity between the allocation occurring under each of the decentralised management scenarios and the Social Optimum will be proportional to the value of time profile externality that occurs in each of these scenarios. In the simulation above the residential sector's preferred profile deviates further from the socially optimal allocation than that of agriculture and would be less preferred as a second best management option. Indeed, following Provencher and Burt (1993),[9] the externality with residential management with a groundwater stock of 70 units is 0.44, whilst the externality with agricultural management is 0.04. This illustrates the benefits to be received from vesting property rights in sectors which better represent social preferences, in this case agriculture.[10]

5. Optimal Groundwater Management and Redundant Reservoirs

Where an externality exists such as the time profile externality described above, there is an incentive for the affected third parties to engage in second-best mitigating or averting behaviour. In the context of the hydrological model described above an incentive exists for the residential sector to mitigate against the additional risks imposed upon them through the actions of the agricultural sector by investing in storage capacity. This section analyses how groundwater allocations under central and decentralised control determine the incentives for the investment in additional storage capacity and how the internalisation of the time-profile externality described above can ameliorate the need for mitigating investment, potentially making reservoirs redundant.

5.1. Central Control with Two Instruments for Storing Water

If the central planner has the option of investing in water storage capacity in order to smooth the flow of water from upstream across seasons for the residential sector, the problem described in the previous sections can be augmented. It is assumed that the residential sector can save an amount of water, K, from period 0 (the wet season), for use in period 1 (the dry season), at a cost reflected by the function $\phi(K)$.[11] The Lagrangian describing the central control problem is:

$$\begin{aligned} L = \max_{u_0, K} & [\alpha[h^a(u_0^a) - c(x_0)u_0^a - \rho^a(u_0^a, Z)] \\ & + (1-\alpha)[h^r(w_0) - \rho^r(w_0, Z_k)] - \phi(K)] \\ & + \beta E\{\alpha[h^a(u_1^a(Z)) - c(Z)u_1^a(Z)] \\ & + (1-\alpha)h^r(w_1)\} + \lambda_0(x_0 - u_0^a), \end{aligned} \tag{11}$$

where

$$w_0 = u_0^r - K = u_0^a - K, \tag{12}$$

$$w_1 = u_1^r + f(K) = u_1^a + f(K). \tag{13}$$

Z_k refers to the stock of water available to the residential sector in period 1; $Z_k = Z + f(K)$, which is an argument in the risk premium for the residential sector.[12] As before it is assumed that the stochastic surface water flow will be used before the certain flow of w_1 is used. The solution to the central control problem is characterised by the following first order conditions. For the socially optimal allocation of groundwater:[13]

$$\begin{aligned} u_0^*: \quad & \alpha\left[\frac{\partial h^a}{\partial u_0^a} - c(x_0) - \frac{\partial \rho^a}{\partial u_0^a}\right] + (1-\alpha)\left[\frac{\partial h^r}{\partial w_0} - \frac{\partial \rho^r}{\partial w_0}\right] \\ & = +\alpha \cdot \beta E\left\{\left[\frac{\partial h^a}{\partial u_1^a} - c(Z)\right]\frac{\partial u_1^a}{\partial Z}\right\} + (1-\alpha)\cdot\beta E\left\{\left[\frac{\partial h^r}{\partial w_1}\right]\frac{\partial u_1^a}{\partial Z}\right\} \\ & - \alpha\cdot\beta E\left\{\left[\frac{\partial c(Z)}{\partial Z}\cdot u_1^a(Z)\right]\right\} - \left(\alpha\cdot\frac{\partial \rho^a}{\partial Z} + (1-\alpha)\cdot\frac{\partial \rho^r}{\partial Z_k}\right). \end{aligned} \tag{14}$$

Again we notice the three terms on the right hand side reflect the stock, pumping and risk components of marginal groundwater user cost. In the two-instrument case we have the additional first order condition regarding the socially optimal level of investment in storage capital:

$$(1-\alpha)\left[\frac{\partial h^r}{\partial w_0}\cdot\frac{\partial w_0}{\partial K}-\frac{\partial \rho^r}{\partial Z_k}\cdot\frac{\partial Z_k}{\partial K}-\frac{\partial \rho^r}{\partial w_0}\cdot\frac{\partial w_0}{\partial K}-\phi'(K)\right]$$
$$+\beta E(1-\alpha)\left\{\frac{\partial h^r}{\partial w_1}\cdot\frac{\partial w_1}{\partial K}\right\}=0. \tag{15}$$

Some algebra gives the more intuitive (16):[14]

$$\left[\beta E\left\{\frac{\partial h^r}{\partial w_1}\cdot\frac{\partial w_1}{\partial K}\right\}-\left[\frac{\partial \rho^r}{\partial Z_k}-\frac{\partial \rho_r}{\partial w_0}\right]-\frac{\partial h^r}{\partial w_0}\right]=\phi'(K). \tag{16}$$

Rearranging (15) and inserting (16) provides (17):

$$K^*:\quad \alpha\left[\left[\frac{\partial h^a}{\partial u_0^a}-c(x_0)-\frac{\partial \rho^a}{\partial u_0^a}\right]+\frac{\partial \rho^a}{\partial Z}\right.$$
$$\left.-\beta E\left\{\left[\frac{\partial h^a}{\partial u_1^a}-c(Z)\right]\frac{\partial u_1^a}{\partial Z}-\frac{\partial c(Z)}{\partial Z}\cdot u_1^a(Z)\right\}\right]$$
$$=(1-\alpha)[\phi'(K)-\eta]. \tag{17}$$

Equation (17) states that the investment in water storage should be undertaken until the marginal benefit of transferring water from the wet period 0 to the dry period 1 for the residential sector is equal to the marginal cost of investment.[15] This relationship arises through the conjunctive nature of the water resource and highlights the trade-off the social planner faces between the two modes of water storage: groundwater management or capital investment. This will be explained in more depth below.

5.2. Decentralised Control with Two Instruments for Water Storage

In the decentralised control scenario the two sectors behave independently and maximise their own expected utilities over the two periods through their respective choice variables: for the agricultural sector and K for the residential sector. Again property rights to the groundwater resource are well defined and vested in the agricultural sector. The sectors are linked only by the conjunctive nature of groundwater and surface water, and as before the time profile of groundwater use by the residential sector is dictated by the agricultural sector. Under these conditions the decentralised outcome is characterised by the following first order conditions.

$$u_0^{a*}: \quad \left[\frac{\partial h^a}{\partial u_0^a} - c(x_0) - \frac{\partial \rho^a}{\partial u_0^a}\right]$$

$$= \beta E\left\{\left[\frac{\partial h^a}{\partial u_1^a} - c(Z)\right]\frac{\partial u_1^a}{\partial Z}\right\} - \beta E\left\{\left[\frac{\partial c(Z)}{\partial Z}\cdot u_1^a(Z)\right]\right\} - \frac{\partial \rho^a}{\partial Z}, \quad (18)$$

$$K^{r*}: \quad \left[\beta E\left\{\frac{\partial h^r}{\partial w_1}\cdot\frac{\partial w_1}{\partial K}\right\} - \left[\frac{\partial \rho^r}{\partial Z_k} - \frac{\partial \rho_r}{\partial w_0}\right] - \frac{\partial h^r}{\partial w_0}\right] = \phi'(K). \quad (19)$$

Equation (18) reflects the fact that the agricultural sector will allocate groundwater to equate its expected marginal benefits with its private marginal user cost, with no regard for the residential sector. Equation (19) reflects how the residential sector manages water-related risks, given the behaviour of the agricultural sector, through the use of its own instrument K.[16]

5.3. Analysis of the Two Instrument Case – Missing Markets and Redundant Reservoirs

The issues of interest in the two-instrument scenario are the incentives for sub-optimal investments in smoothing capital due to presence of the time profile externality, and the extent to which optimal groundwater management might affect the timing and scaling of investments in water storage.

5.3.1. *Decentralised Investment in Water Storage – Missing Markets*

Equation (19) is a sufficient condition to determine the optimal investment decision in the decentralised scenario. The first term on the left hand side represents the expected marginal return from non-stochastic additions to the water supply in period 1. The second term (in brackets) represents the way in which the risk premium changes as a result of the transfer of water across time. The last term represents the marginal benefit from additional units of water in period 0. In sum left hand side represents the net marginal changes in expected utility that arise through the transfer of water across time and is the difference between the marginal benefit of non-stochastic water in period 1 and period 0.[17]

As seen in Section 3, this disparity may arise as a result of a time profile of water that is determined externally to the residential sector by a sector with different preferences, i.e. on account of missing markets. If another sector is in control of the groundwater stock then the residential sector is only able to use the instrument of capital investment to manage the residual stochastic surface flow risk. It does this by investing up until the point at which the marginal benefit of water transfer is equated to $\phi'(K)$, the marginal cost of capital function. In this way the residential sector mitigates against the time profile externality.

RESULT (Capital Investments under Missing Markets). Investment in water storage by the residential sector may occur in the decentralised scenario as the sole means available for meeting that sector's preferences for water smoothing, and in response to other sector's failure to take those preferences into consideration.

$$K^{r*} > 0 \quad \text{if} \quad \beta E\left\{\frac{\partial h^r}{\partial w_1} \cdot \frac{\partial w_1}{\partial K}\right\} - \frac{\partial \rho^r}{\partial Z_k} > \frac{\partial h^r}{\partial w_0} - \frac{\partial \rho_r}{\partial w_0}. \tag{20}$$

Proof. Restatement of Equation (19). □

In addition, it is more likely that such investments will occur if the two sectors are more heterogeneous. This may be seen from the fact that the incentive to invest will increase as the difference in (22) increases. Under the assumptions concerning functional forms,[18] and noting that a fixed stock of groundwater means that a decrease (increase) in leads to an increase (decrease) in , then a decrease in will cause this difference to increase (for a risk averse downstream residential sector).[19] Thus the disparity increases with increasing heterogeneity of preferences regarding the time profile of use. In order to satisfy the relevant first order conditions a higher (lower) level of investment in water storage is required, as the two sectors become more (less) heterogeneous.

5.3.2. *Socially Optimal Investment in Water Storage – Redundant Reservoirs*

The socially optimal investment in water storage is characterised by Equations (16) and (17). It can be seen that (16) is formally equivalent to (19) and the same reasoning applies as described above. Equation (17) characterises how optimal decisions will be made concerning the two modes of water storage available to the social planner: groundwater management and/or capital investment. In the Social Optimum, these management/investment decisions are made simultaneously, rather than in the segregated manner of the decentralised problem. The left hand side of (17) reflects the net social costs to the agricultural sector of transferring water from period 0 for period 1.[20] The social cost of investing in water storage capital is reflected by the right hand side of (17).

In essence, the cost of using groundwater management as an instrument for water transference are those associated with altering the time profile of water use upstream towards that preferred by the downstream sector. At the optimum these costs are weighed against the costs of investing in water storage. Equation (17) implies that where the marginal social cost of altering the time profile for agriculture is less than the marginal social cost of investment, groundwater is the efficient smoothing device for the residential sector. Where investment is less costly this is the efficient smoothing device.[21] In this way a socially optimal trade-off occurs between the two storage modes in determining the optimal time-profile of water, and the time-profile externality is internalised.

It is evident that one solution to (15), (16) and (17) is where there is zero capital investment. This would occur where the marginal benefits are equated to the marginal user cost of groundwater for each individual sector.[22] Again, the disparity between the decentralised and the centralised management regimes hinges upon the extent of the heterogeneity of preferences and thereafter the social weights attached to each sector.

Where the heterogeneity is such that an incentive exists for the residential sector to make mitigating investments, i.e. where the residential sector is more risk averse than agricultural sector, groundwater allocation will be able to attenuate some or all of the difference found in (20). Therefore the optimal capital investment (i.e. with optimal groundwater management) will always be less than the level of investment occurring under decentralised management, and the difference between the cost of investment in each case measures the time profile externality. This is simply because the use of both instruments together affords the social planner more flexibility than using each in isolation.

RESULT (Optimal Storage Capital Investment). Where capital investment is indicated, socially optimal level of capital investment will always be less than that which would occur in the decentralised scenario:

$$K^{**} < K^{r*} : \text{ if}$$

$$\beta E\left\{\frac{\partial h^r}{\partial w_1}\cdot\frac{\partial w_1}{\partial K}\right\} - \frac{\partial \rho^r}{\partial Z_k} > \frac{\partial h^r}{\partial w_0} - \frac{\partial \rho^r}{\partial w_0}, \quad \text{at the optimum.} \qquad (21)$$

Proof. See Appendix B. □

In other words the optimal control of groundwater can realise the buffer values of the resource rather than having these values capitalised in investments into dams. To the extent that such forms of externalities exist and the markets for their attenuation fail, some dams are only the consequence of inefficient water management policies.

6. Conclusion

Why are there so many dams on the world's waterways? This paper has argued that one explanation for the existence of some dams is that inefficient water management policies fail to apply existing water resources to the necessary purposes. When water management policies fails to take into consideration the needs and preferences of all of the waterway's users, then those users must makes use of the instruments at their disposal for the solution of their problems. One of the instruments at the disposal of downstream water users is the construction of river dams. This instrument transfers water from a season of water surplus to one of relative deficit, through the creation of an artificial reservoir.

Of course such reservoirs often exist naturally in the vicinity of a watercourse, beneath the surface or above the catchment area concerned. The problem is that usually the water users concerned have no property rights in the natural reservoir, and so they must create their own artificial one. This is inefficient if the social costs of the artificial reservoir (construction and river blockage) exceed the social costs of re-allocating the natural reservoir (change in its time of use). When it is inefficient to create the artificial reservoir rather than manage the natural one, it is a redundant reservoir.

This chapter has examined and explained the simplest case of the natural reservoir that could potentially serve two sectors costlessly, and where capital investments in artificial reservoirs involve only their construction costs. Of course, the more complicated cases concern natural water resources over which there are many conflicts, and dams that create many categories of costliness. However, the general points we wish to emphasise are, firstly, that the water resource is one for which market failures are replete (even when property rights are well-defined in many dimensions) and, secondly, inefficient water management policies induce the specific forms that some of the inefficiencies might take. Markets are missing, and hence some reservoirs are redundant.

Appendix A

The central control problem is characterised by the Lagrangian derived from the problem statement:

$$\begin{aligned} V_0(x_0) &= \max[\alpha[h^a(a_0^a - c(x_0)u_0^a - \rho^a(u_0^a, Z)] + (1-\alpha)[h^3(u_0^r) - \rho_r(u_0^r, Z)]] \\ &\quad + \beta E\{\alpha[h^a(u_1^a(Z)) - c(Z)u_1^a(Z)] + (1-\alpha)h^r(u_1^r(Z)\} \\ &\quad \text{S.T.} \quad Z = x_0 - u_0^a \geq 0. \end{aligned}$$

The solution satisfy the first order conditions and the Kuhn–Tucker conditions (A.2) and (A.3). Following Provencher and Burt (1993), the first order conditions come from:

$$\begin{aligned} &\alpha\left[\frac{\partial h^a}{\partial u_0^a} - c(x_0) - \frac{\partial \rho^a}{\partial u_0^a} + \frac{\partial \rho^a}{\partial Z}\right] + (1-\alpha)\left[\frac{\partial h^r}{\partial u_0^a} - \frac{\partial \rho^r}{\partial u_0^a} + \frac{\partial \rho^r}{\partial Z}\right] \\ &= \beta E\left\{\frac{\partial V_1}{\partial Z_0}\right\}, \end{aligned} \tag{A.1}$$

where

$$\beta E\left\{\frac{\partial V_1}{\partial Z_0}\right\} = C_s^* + C_D^*.$$

As defined in Section 3.1

$$x_0 - u_0^a \geq 0, \quad \lambda_0 \geq 0, \tag{A.2}$$

$$\lambda_0[x_0 - u_0^a] = 0. \tag{A.3}$$

Appendix B

Optimal control of groundwater reduces storage investment. The proof is as follows:

(i) Central Control: rewriting Equation (14) assuming equal social weights as follows:

$$\mathrm{MB}_0^a + \mathrm{MB}_0^r = \mathrm{MUC}_1^a + \mathrm{MUC}_1^r, \tag{B.1}$$

where MUC^i refers to the expected marginal user cost of groundwater to sector i, and MB^i refers to the net marginal benefit of water use in period 0 for sector i. It follows that Equation (16) can be written as:

$$\mathrm{MUB}_1^r - \mathrm{MB}_0^r = \phi'(K^{**}). \tag{B.2}$$

Rearranging Equation (B.1) to obtain:

$$\mathrm{MB}_0^a - \mathrm{MUC}_1^a = \mathrm{MUC}_1^r - \mathrm{MB}_0^r \tag{B.3}$$

and using (B.2) we obtain:

$$\mathrm{MB}^a - \mathrm{MUC}_1^a = \mathrm{MUC}_1^r - \mathrm{MB}_0^r = \phi'(K^{**}), \tag{B.4}$$

which mirrors in general terms Equation (17). Supposing that the sectors have heterogeneous preferences such that the solution to (B.1)–(B.3) is *not*:

$$\mathrm{MB}_0^a - \mathrm{MUC}_1^a = \mathrm{MUC}_1^r - \mathrm{MB}_0^r = 0$$

and the heterogeneity is such that there is a positive level of investment required at the optimum, i.e. (B.2) holds with positive K: $\phi'(K^{**}) > 0 \Rightarrow K^{**} > 0$. Given (B.3) and (B.4) this implies that $\mathrm{MB}_0^a - \mathrm{MUC}_1^a > 0$, i.e. the disparity of inter-temporal marginal benefits and marginal user costs is equal and opposite for each sector.

(ii) Decentralised Management: Re-writing (18) in the same way, using lower case characters to differentiate the decentralised problem from the central we obtain:

$$\mathrm{mb}_0^a = \mathrm{muc}_1^a, \tag{B.5}$$

which characterises the decentralised groundwater allocation. Similarly we can write:

$$\text{muc}_1^r - \text{mb}_0^r = \phi'(K^{r*}). \tag{B.6}$$

which again characterises the investment decision and is formally equivalent to (B.2).

(iii) Comparison of Central and Decentralised Management: Comparing the groundwater allocations in the two management scenarios it is easy to see that where investment is positive, and preferences are heterogeneous, in the social optimum for the agricultural sector it is the case that $\text{MB}_0^a - \text{MUC}_1^a > 0$, whilst in the decentralised problem $\text{mb}_0^a - \text{muc}_1^a = 0$. Given the same social weights, preferences and groundwater stock in each management scenario, these conditions are synonymous with a higher level of water use in the first period in the decentralised case: $U_0^* < u_0^{a*}$.[23] This in turn means that a higher level of investment will be necessary in the decentralised problem in order to satisfy (B.6) (or (19)). More formally:

$$\text{muc}_1^r - \text{mb}_0^r > \phi'(K^{**}) \tag{B.7}$$

and therefore from (B.2) and (B.7),

$$\phi'(K^{r*}) > \phi'(K^{**}) \Rightarrow K^{r*} > K^{**}. \tag{B.8}$$

Therefore where preferences differ in a way that gives rise to incentives for investment in water storage for the residential sector, the investment in the centralised management scenario will always be less than or equal to that which occurs in the decentralised scenario. I.e. the social planner will never choose to invest more than the residential sector would in the decentralised scenario.

Notes

[1] The benefit function $h(u, q)$ is assumed to be concave in water inputs ($h' > 0, h'' < 0$).

[2] It will be seen later that stochasticity of surface water flows is not a necessary condition for the time profile externality to exist, however in order to illustrate how the time profile externality relates to the taxonomy of Provencher and Burt (1993) the stochastic assumption is maintained from hereon.

[3] Of course there are other preference-based means of generating heterogeneity in preferred time profiles of use (e.g. different preferences for water smoothing generally). We use the risk premia and stochastic surface water approach merely as a case in point.

[4] A recent paper by Charles Howe has noted that in Colorado the time profile of use by land owners is now coming under scrutiny and possible regulation in that jurisdiction (Howe, 2000).

[5] Of course, if $\delta = 0$, then use of groundwater by the upstream user has no impact on the downstream user's water availability, and no time of use externality considerations apply. For all

other values of δ, there is a potential externality over time of use. We choose $\delta = 1$ in order to focus on the time of use externality to the exclusion of all other externalities.

[6] Remembering the assumption that $\delta = 1$, which enables the upstream sector to determine the time of use, rather than the use itself.

[7] The first two terms on the right hand side of (7) are equal to $\beta E\{\partial v_1/\partial Z\}$, see Appendix A. The last term in Equation (7) reflects the combined risk costs of the two sectors, which in turn reflect their valuation of how the riskiness of incomes in the terminal period varies with the groundwater stock, Z.

[8] λ_0 is assumed to equal zero in this formulation, indicating that the stock is not exhausted in the period 0.

[9] We follow their method by measuring the externality as the difference between the marginal user cost in the social optimum and each of the decentralised options.

[10] One reason for agriculture more closely representing the social optimum is that it bears the social cost of the extraction of groundwater in this example.

[11] The formulation assumes that investment occurs at the start of period 0 in the both scenarios. At this point $\phi(K)$ is assumed to be a non-negative convex, monotonic function of K.

[12] $f(K)$ reflects that additions to the stock of water for period 1 are not necessarily used in full at that time.

[13] We assume throughout that λ_0 is equal to zero. This is reflects that the groundwater stock is not exhausted in period 0.

[14] Noting that from (12) $\partial w_0/\partial K = -1$, and $\partial Z_k/\partial K = 1$, and dividing through by the social weight $(1 - \alpha)$.

[15] Less a term η reflecting the different ways in which the different stocks (groundwater and the water storage investment) contribute to w_1. Note

$$\eta = \beta E\left\{\frac{\partial h^r}{\partial w_1}\left[\frac{\partial w_1}{\partial K} - \frac{\partial w_1}{\partial Z}\right]\right\},$$

where $\partial w_1/\partial z = \partial u_1^a/\partial Z$. It is assumed that $\eta = 0$.

[16] It is worth noting that both (16) and (19) are independent of the social weights. This is largely because the decision to invest does not directly affect the upstream agricultural sector.

[17] Expected Utility should be thought of in terms of Equation (4) above.

[18] For this statement to be unambiguous the risk premium must increase with reduced levels of non-stochastic water in the stochastic period 1; the marginal effect of Z_k is not outweighed by the marginal effect of w_0 (wealth).

[19] Given K is fixed in the short term, and changes in w_0 come from u_0.

[20] The mechanics of this statement are as follows. As in Section 3, the social optimum reflects a groundwater allocation that is a weighted average allocation of the preferences of the respective sectors. Therefore should the conditions exist for a positive downstream investment; the expected marginal benefit of water in period 1 is greater than that in period 0 for the residential sector, then given (15) it must be the case that the opposite relationship holds for the agricultural sector.

[21] Note that where the right hand side of (17) is negative we are dealing with a situation in which the agricultural sector would prefer a transfer of water from period 0 to period 1 and the residential sector would prefer the opposite transfer. In which case investment in a dam is not an option in the 2 period model.

[22] Given the assumptions regarding $\phi(K)$.

[23] Remembering that lower u_0^a results in a lower marginal user cost for groundwater.

References

Allen, R.C. and Gisser, M. (1984) Competition versus optimal control in groundwater pumping when demand is non-linear, *Water Resources Research* **20**(7), 752–756.

Boronina, A., Cristodoulides, A., Renard, P. and Balderer, W. (2001) Hydrogeological aspects of the water supply in Cyprus: KOURIS catchment, Chapter 4 of Cyprus Integrated Water Resource Management Project, CSERGE, UCL.

Bredehoeft, J.D. and Young, R.A. (1983) Conjunctive use of groundwater and surface water for irrigated agriculture: Risk aversion, *Water Resources Research* **19**, 1111–1121.

Burt, O. (1967) The temporal allocation of groundwater, *Water Resources Research* **3**, 45–56.

Cummings, R.G. (1969) Some extensions of the economic theory of exhaustible resources, *Western Economic Journal* **8**(3), 209–220.

Cummings, R.G. and Winkelman, D.L. (1970) Water resource management in arid environs, *Water Resources Research* **6**, 1559–1560.

Dasgupta, P. (1982) *The Control of Resources*, Cambridge University Press, Cambridge, Chapter 6.

Gisser, M. and Sanchez, D.A. (1980) Competition versus optimal control in groundwater pumping, *Water Resources Research* **31**, 638–642.

Hart, O. and Moore, J. (1990) Property rights and the nature of the firm, *Journal of Political Economy* **98**, 1119–1158.

Howe, C. (2000) The management of renewable and non-renewable groundwater resources with observations on the special problems of island states and coastal areas, Paper presented to the Cyprus Water Resource Management Symposium, Nicosia, September 2000.

Krutilla, J. and Fisher, A. (1967) Conservation reconsidered, *American Economic Review* **57**(4), 777–786.

MITWRG (1999) Solutions to water scarcity in the Republic of Cyprus – A proposal for water banking, The Massachusetts Institute of Technology Water Resources Group.

Ostrom, E. (1995) Designing complexity to govern complexity', in S. Hanna and M. Munashinghe (eds.), *Property Rights and the Environment – Social and Ecological Issues*, The Beijer International Institute of Ecological Economics and the World Bank, Chapter 3, pp. 33–47.

Provencher, B. (1994) Conjunctive-use of surface water and groundwater, in D.W. Bromley (ed.), *The Handbook of Environmental Economics*, Blackwell, Oxford, pp. 503–521.

Provencher, B and Burt, O. (1993) The externalities associated with common property exploitation of groundwater, *Journal of Environmental Economics and Management* **24**, 139–158.

Tsur, Y. (1990) The stabilisation role of groundwater when surface water supplies are uncertain: The implications for groundwater development, *Water Resources Research* **26**, 811, 818.

Tsur, Y. and Graham-Tomasi, T. (1991) The buffer value of groundwater with stochastic surface water supplies, *Journal of Environmental Economics and Management* **21**, 201–224.

World Commission on Dams (2000) Report on the economic consequences of dams on natural waterways, WCD, Capetown.

4.2. *Water as an Industrial Management Problem – Vertical Considerations*

Regulation of Public Utilities under Asymmetric Information: The Case of Municipal Water Supply in France

SERGE GARCIA and ALBAN THOMAS

1. Introduction

In France, contrary to other sectors such as electricity and telecommunications, the policy decision-makers have not yet included recent recommendations from contract theory in setting up regulation systems and authorities for water utilities. Environmental regulation for water quality is one of the attributions of the Water Agencies created in 1964, that control for industrial and local community pollution of water through a system of effluent emission taxes and pollution abatement subsidies. The legislative trend on environment since the early 1990s is clearly characterized by more and more stringent regulation on effluent emissions and on privately or publicly-owned abatement units. Yet public authorities do not exist that may set up rules based on economic principles that should be enforced by private companies in charge of water public services.

When a local community decides to delegate the provision of drinking water supply services to a private operator, it is within the framework of a medium-run contract defining the rules that the firm has to follow, in terms of expected provision of services, water rate charged to the customers and so on. In a majority of cases, the initial capital expenditures are initiated by the local community itself, while the leaseholder has the obligation to operate the water utility and to conduct maintenance works whenever appropriate. The contract between the local community and the private firm is therefore the most important way through which optimal allocations of water resources can be implemented.

Since the beginning of the 1980s, research in contract theory has studied problems of asymmetric information in modeling the relationship between a public regulator and a private monopoly. It is assumed that a strategic advantage can be present for the regulated firm because of a private information parameter characterizing its technology. The contracts considered typically induce the firms to reveal their private information by using a revelating mechanism. Two major streams of research attempt to explain the regulation process along these lines, when production costs of the firm are not observed (Baron and Myerson, 1982), or when an accounting audit of the firm allows to use information on the ex-post firm costs (Laffont and Tirole, 1986).

P. Pashardes et al. (eds.),
Current Issues in the Economics of Water Resource Management, 195–217.

However, empirical applications of these models (in particular, the relationship between a regulator and a private monopoly by way of contracts) are still very few. In most econometric works dealing with water industry (Mann and Mikesell, 1976; Crain and Zardkoohi, 1978; Hayes, 1987; Kim, 1987; Renzetti, 1992b), the implications of asymmetric information are ignored, and only a few studies (Wolak, 1994; Thomas, 1995; Lavergne and Thomas, 1997) have modeled these problems. Other works on other industries account for the asymmetry of information (Dalen and Gomez-Lobo, 1996, 1997; Gagnepain and Ivaldi, 1999 in the urban transportation). Auriol et al. (1999) show the advantage of allowing for the presence of imperfect information case by a non-nested test in the US local telephone market.

For water industry, an important feature is the occurrence of network leaks. This loss is clearly a concern for water utility managers in terms of opportunity costs, all the more so since actual losses are 25% on average and can reach 50% of the distributed water volume. From a public viewpoint, water losses are undesirable in a context of resource conservation policy, especially in areas where shortage of water can be severe.

The existence of network leaks has not been fully investigated in the literature on water utilities. Water utility managers seeking to reduce inefficiencies would clearly account for such an interaction in their production plans. Besides, federal or local environmental and resource regulation authorities may need to address this issue when designing performance-based regulatory schemes. The utility manager may find profitable not to minimize water network losses, in particular by not repairing all leaks (see Garcia and Thomas, 2001). If the utility manager decides not to repair because of the significant cost of doing so, there will be no cost saving on water production. Nevertheless, the manager may prefer this latter solution because increasing water output supply is less costly. If the problem is to satisfy an increasing customer demand, the adopted solution may be to produce more water instead of repairing network leaks. Hence, because repairs are highly labor-intensive while over-producing increases mostly the cost of energy, joint production of a water volume delivered to customers and a lost water volume will be preferred if the marginal cost of labor is higher than the marginal cost of electricity.

We may think that cost arguments are not the only ones to justify such level of losses. There maybe exists an informational reason for which the local community requires the operator not to reduce this level.

The purpose of this chapter is to examine the impact of asymmetric information on the production decision of regulated water public utilities. Before delegating operation, the local community faces two problems: when it has to choose the private operator, it does not know the efficiency of the firms which respond to competitive bidding. Moreover, the effort that the operator will exert to improve the network quality by reducing the water losses is not clear, in particular because cost-reducing effort is obviously expensive (see arguments above). The regulation is based on a contract in which produced and delivered water and network losses volumes are observable. From the model in which we suppose that water losses are socially costly, we first derive optimal solutions in the case of perfect information. The second case in which the operator has a private information on its technology (θ) is then examined. The empirical analysis consists of estimating parameters of interest on demand and technology (in particular θ) and to use them to simulate the first and second-best contract solutions.

The remainder of the chapter proceeds as follows. The organization of water industry in France is briefly described in Section 2. Section 3 presents the technology of municipal water supply networks and Section 4 explores the demand side of the water sector and stresses the problem of losses for customers. The economic model of regulation and the optimal contracting design in the case of asymmetric information are presented in Section 5. Section 6 is the empirical application to the French water industry. Estimation of cost and demand parameters, and of the private information parameter distribution, allows to simulate the optimal contract solutions. Section 7 is the conclusion.

2. The French Water Supply Industry

The water industry shares common features with other network industries (electric power, telephone, urban transport). In particular, conditions for natural monopoly prevail in this sector, where water distribution is essentially local because of high water transportation costs. Furthermore, important fixed and sunk costs constitute a real barrier to entry to these local services. The chief objective of water services is to produce water with sufficient quality from a resource (groundwater or surface water) that may necessitate preliminary treatments to make water safe and palatable, and to distribute water by continuously adaptating supply with daily demand while preserving water quality during its transportation in the transmission pipelines and distribution mains.

Concerning water quality, there exist many standards both from national and European legislation. Two important European Commission directives deal with the quality of surface water for the production of drinkable water (June 16, 1975), and with the quality of water for human consumption (July 15, 1980). The latter directive defines maximum concentrations for 65 parameters of drinking water quality. At a national level, regional Water Agencies in France have adopted a resource conservation policy based on effluent emission and extraction charges for industrials and local communities. Tax revenues are redistributed among polluters in the form of financial assistance for investing in pollution abatement or resource management facilities. At the same time, European directives regarding wastewater effluent emissions are translated into national legislation (the French Water Act of 1992), which imposes uniform effluent emission and water quality standards to all local communities.

In France, local communities are responsible for water services delivered to local customers (drinking-water delivery and wastewater treatment). Direct (public) operation of water services by local communities may take several forms: direct management, management with financial autonomy and/or moral authority. The other possibility is to delegate those services to a private company, along alternative contracting schemes with the local community: concession, lease contract, management contract, commissioner management contract. There are no central regulation authorities to set the rules in management or tarification for the services. The rules that private operators have to respect are defined in a multi-year contract.

Contracts typically specify the nature of expected services, the price of water and the rules of price updating and, in the case of lease contracts, the part which is paid back by the operator to the community for its investments. Moreover, local communities are always the owners of equipment (pipelines, treatment and stocking facilities, etc.) even in the case when they delegate operation and possibly investment in new or additional facilities to the private firm. Leasing is the most common form of contract; contract length can vary from 7 to 12 years. The private firm has the responsability for operation and maintenance of the water utility, it collects tariff revenues from the users and pays a special additional charge to the local community, which is included in the water rate determined by the contract. The maximum basic rate is determined by the contract from projected financial accounts provided by the operator. Periodic revisions of water rate are defined using a price index updating formula. In France, the effective tariff is in most cases a two-part tariff corresponding to the provision of water supply and wastewater services. For the two activities, the local community and the operator set fixed charges and a price per unit of water consumed. Various fees and taxes are added to these prices as Water Agency fees, National Fund for the Development of Water Supply Systems fees and Value Added Tax.

3. Technology and Cost of Production

Water supply results from two successive stages which are respectively water production and its delivery to final users (customers). The production process involves extraction of raw water (this volume is called the extracted volume V_e) and its treatment if it is unsuitable for consumption. Water is then pressurized to make it circulate in the delivery network.[1]

The delivery activity has the objective to satisfy the demand by delivering directly at customers' taps and charging a given water quantity level. We denote V_c the level of water consumption (in cubic meter). Because a part of distributed water volume does not reach its goal, we have $V_e > V_c$ and $V_e = V_c + V_l$, with V_l the water losses in the delivery network. Trivially, only two out of the three water volumes are sufficient to describe production. We choose to model the technology with the consumed volume V_c and the lost volume V_l.

In the remainder of the chapter, we will denote Y_c the consumed volume per customer (in cubic meter) by dividing the consumed volume by the number of customers, denoted n. Y_l represents the network losses per customer which, as explained above, are endogeneous. As is common in the water industry, we consider a measure of loss called linear index of losses (I) which is the ratio of lost volume per customer to the length (*Leng*) of transmission pipelines and distribution mains, that is: $I = Y_l/Leng$. This index relates the volume of losses to the importance of the network and hence is useful as a proxy for the physical state of the network.

Actual costs of the privately-operated water utility are not observed. Water utilities have to present an annual financial report of their activities, but accounts in this report may aggregate expenses that are not related to current operation, hence may not be precise. We assume that the water utility minimizes its operating expenses given the existing capital and the technology of production. Hence, under the usual conditions of regularity (see Lau, 1976), there exists a variable cost function that we can write as:

$$C = C(V_c, I, w, \theta),$$

where w is the vector of variable input prices (labor, electricity and a third category including material, repairing and maintenance works). $\theta > 0$ characterizes the efficiency of the utility. We assume θ is defined on the interval $[\underline{\theta}, \bar{\theta}]$, with $\underline{\theta}$ the more efficient utility and $\bar{\theta}$ the less efficient utility. As said above, there is a level of effort to reduce water losses, which represents the quality of network. Effort for reducing network losses is costly, and we suppose that the disutility of effort (investment costs of leaks repairing) is included in the total variable cost of supplying water. Hence, increasing effort level (i.e., the quality of network, all other things being equal), first reduces and then increases costs, when the value of leaks reduction becomes less than the cost of effort. This effort is observable by the way of V_c and I. Total cost to the private operator is the sum of variable and fixed costs of production: $C + w_K K$, where K is capital stock (here represented by the length of network), with w_K the unit cost of capital.

For the setting-up of the contract model, we have to make the following assumptions about the cost function (the underscript denotes partial differentiation): $C_\theta > 0$ means that a utility with a high θ (inefficient) produce with a higher cost. $C_{\theta I} < 0$: it is more costly for the less efficient utilities to decrease the lost volume. This means that all other things being equal, improving the quality of network is more difficult for these utilities.

4. Customer Surplus and Demand

The gross surplus of consumers is given as $nS(Y_c)$, where n is the number of consumers. It is assumed positive. S represents the welfare function depending on the consumption level of drinkable water, with $S' > 0$, $S'' < 0$ and $S(0) = 0$. We denote $P(Y_c) = \partial S(Y_c)/\partial Y_c$ the inverse demand function of consumers. If we denote by T the fixed fee that the user pays for the water supply service, the net surplus of consumers is written as: $n[S(Y_c) - P(Y_c)Y_c - T]$. With assumptions given above, the net surplus is concave in its arguments.

We then introduce an additional cost, borne by the community and independent of consumption, because users are more or less directly affected by the resource extraction which is only semi-renewable and the network losses, for the following reasons:

- First, they are directly affected by the waste of water resources, that may involve a decrease in raw water stock and therefore a possible rationing for them in the near future (in particular in periods of drought).
- Second, insufficient water pressure may occur at the tap of the user, due to leaks in the water network pipelines. And in the extreme case, leaks or breakings of pipelines can interrupt the provision of water.
- Third, the indirect consequences of leaks on the users can be various. We can note the possibility of flooding, road traffic interruptions and others pertubations on the other public services, and damages for the users at home.

$D(V_c, I)$ is the damage induced by resource extraction and network water losses, such that $D_{V_c} > 0$, $D_{V_c V_c} > 0$. We have the similar impact of I on the damage: $D_I > 0$, $D_{II} > 0$, i.e., losses increase damage at an increasing rate. Finally, $D_{IV_c} < 0$ is negative: the more water extracted the less important the marginal damage with respect to I. Hence, the gross surplus trivially depends on volume consumed by the users of service but also on extracted water quantities.

5. The Optimal Contract

In nearly all delegation contracts of public service to a private firm, almost all of fixed costs are supported by the local community. In water supply (contrary to wastewater treatment), the stock of capital is typically operated for a long time period and asset depreciation is already significant. Besides, a part of maintenance and operating works (essentially replacement) is assigned to the private operator, and replacement work expenses are included in the operating costs of the operator. Hence, there is a monetary transfer T from the local community to the operator, corresponding to the fixed charge paid by the customer times the total number of customers, for covering fixed costs. In our framework, there is no social cost of public funds related to the monetary transfer because the water budgets of the communities are separated from the municipal budget. The users pay directly by means of the fixed fee and not by collecting local taxes, and by assumption this does not modify their behavior.

The local community maximizes a weighted sum of the net surplus of customers and the operator's profit, but seeing to it that the weight for the customers is more important. The local community proposes a menu of contracts specifying an extracted water volume, a volume consumed by the final customers, an associated price rate, and a monetary transfer representing the degree of capital investment achieved by the community.

5.1. THE PERFECT-INFORMATION CASE

Under complete information, the local community regulates the private operator (the agent) by specifying the quantities and the monetary transfer, knowing the value of private information θ.

As said before, the objective of the regulator is to maximize social surplus as a weighted sum of the net surplus of customers and the operator's profit. Let π denote a weight representing the community preferences. Setting $\pi > 1/2$, the local community favors customers more than it favors the private operator. The problem of the regulator is:

$$\max_{Y_c, I} W = \pi[nS(Y_c) - D(V_c, I) - n(Y_c P(Y_c) - T)] + (1 - \pi)U \tag{1}$$

under the condition: $U = n(T + Y_c P(Y_c)) - C(V_c, I, w, \theta) - w_K K = u$, where U is the firm's profit. We thus require positive profits for the firm to participate. The problem of the regulator becomes:

$$\max_{Y_c, I} \pi[nS(Y_c) - D(V_c, I) - C(V_c, I, w, \theta) - w_K K].$$

The solution for (Y_c, I) obtains by solving the system:

$$P(Y_c) = C_{V_c}(V_c, I, w, \theta) + D_{V_c}(V_c, I), \tag{2}$$

$$D_I(V_c, I) = -C_I(V_c, I, w, \theta). \tag{3}$$

Equation (2) means that unit price is equal to marginal cost of distributing a water volume V_c to final users plus a term which measures resource scarcity and that we call the shadow price of *in situ* groundwater. This equality is true for all weight (π) given to surplus on the two parts. Equation (3) indicates that at a given water volume V_c, the losses index is chosen efficiently, taking into account that the effort is optimal. The monetary transfer T is determined to ensure that $U = 0$:

$$T = -Y_c P(Y_c) + \frac{C(V_c, I, w, \theta) + w_K K}{n}.$$

5.2. The Imperfect-Information Case

We now consider the case where the operator has private information about its costs, namely the efficiency parameter θ. The local community does not observe θ but has a prior information about it, in the form of a density function f and a cumulative density function F on the domain $[\underline{\theta}, \bar{\theta}]$. The community does not directly observe the effort e neither. The use of a revelation mechanism has for purpose to extract the private information from the private operator.

Hence, the local community submits to the private operator a menu of contracts indexed by θ, $(Y_c(\theta), I(\theta), T(\theta))$. The private operator will select the combination that it prefers according to the type θ it reports. Because an operator of type θ can report any another type, the first stage consists in defining the constraints in terms of the operator's profit. The firm's profit[2] characterized by a (true) type θ and a report $\tilde{\theta}$ is:

$$v(\theta, \tilde{\theta}) = nT(\tilde{\theta}) + nP[Y_c(\tilde{\theta})]Y_c(\tilde{\theta}) - C[V_c(\tilde{\theta}), I(\tilde{\theta}), w, \theta] - w_K K.$$

The contract $(Y_c(\theta), I(\theta), T(\theta))$ must verify the first-order constraint of incentive compatibility (IC1):

$$\forall \theta, \theta' \in [\underline{\theta}, \bar{\theta}], \qquad v(\theta, \theta) \geq v(\theta, \tilde{\theta}).$$

Defining $U(\theta) = v(\theta, \theta)$ and assuming that the mechanism is continuously differentiable, we can write this constraint as:

$$U'(\theta) = -C_\theta(V_c(\theta), I(\theta), w, \theta). \tag{4}$$

In other words, the increase in the operator's profit when θ decreases (the operator is more efficient) is equal to the marginal cost with respect to its efficiency parameter. The second-order condition of incentive compatibility (IC2) is:

$$C_{\theta Y_c}(V_c(\theta), I(\theta), w, \theta)\frac{\partial Y_c}{\partial \theta} + C_{\theta I}(V_c(\theta), I(\theta), w, \theta)\frac{\partial I}{\partial \theta} \leq 0.$$

We ignore the (IC2) constraint in the regulator's program; we check *ex post* that it is verified.

From (4), as the cost is increasing in θ, we can note that U is a decreasing function of θ and the individual rationality constraint (IR) $U(\theta) \geq 0$ is simply:

$$U(\bar{\theta}) = 0.$$

The problem of the regulator can be written as:

$$\max_{Y_c(\theta), I(\theta)} \int_{\underline{\theta}}^{\bar{\theta}} \big[\pi[nS(Y_c(\theta)) - D(V_c(\theta), I(\theta)) - n(P(Y_c(\theta))Y_c(\theta) - T(\theta))] + (1-\pi)U(\theta)\big] f(\theta)\, d\theta, \tag{5}$$

such that:

$$U(\theta) = n(T(\theta) + P(Y_c(\theta))Y_c(\theta)) - C(V_c(\theta), I(\theta), w, \theta) - w_K K, \tag{6}$$

$$U'(\theta) = -C_\theta(V_c(\theta), I(\theta), w, \theta), \tag{7}$$

$$U(\bar{\theta}) = 0. \tag{8}$$

By introducing the various constraints in the formulation of social surplus to be maximized, the problem of the regulator becomes:

$$\begin{aligned}\max_{Y_c(\theta), I(\theta)} \int_{\underline{\theta}}^{\bar{\theta}} \pi\big[& nS(Y_c(\theta)) - D(V_c(\theta), I(\theta)) \\ & - C(V_c(\theta), I(\theta), w, \theta) - w_K K\big] f(\theta)\,\mathrm{d}\theta \\ & - (2\pi - 1) \int_{\underline{\theta}}^{\bar{\theta}} C_\theta(V_c(\theta), I(\theta), w, \theta) F(\theta)\,\mathrm{d}\theta.\end{aligned} \tag{9}$$

We denote respectively Y_c et I the optimal volume sold and the linear loss index in the imperfect-information case. The solution (Y_c, I) obtains by solving the system:

$$\begin{aligned}P(Y_c) &= C_{V_c}(V_c(\theta), I(\theta), w, \theta) + D_{V_c}(V_c(\theta), I(\theta)) \\ &\quad + \frac{2\pi - 1}{\pi}\frac{F(\theta)}{f(\theta)} C_{\theta V_c}(V_c(\theta), I(\theta), w, \theta),\end{aligned} \tag{10}$$

$$\begin{aligned}D_I(V_c(\theta), I(\theta)) &= -C_I(V_c(\theta), I(\theta), w, \theta) \\ &\quad - \frac{2\pi - 1}{\pi}\frac{F(\theta)}{f(\theta)} C_{\theta I}(V_c(\theta), I(\theta), w, \theta).\end{aligned} \tag{11}$$

At the second-best optimum solution, the operator receives a positive information rent. We compute this rent as:

$$U(\theta) = \int_{\theta}^{\bar{\theta}} C_\theta(V_c(\tau), I(\tau), w, \tau)\,\mathrm{d}\tau. \tag{12}$$

Hence we can determine the transfer that the community is granting to the private operator:

$$nT(\theta) = U(\theta) - nP(Y_c(\theta))Y_c(\theta) + C(V_c(\theta), I(\theta), w, \theta) + w_K K.$$

The usual hypothesis of a monotonic hazard rate (i.e., log concavity of F) is satisfied by the non-decreasing nature of $F(\theta)/f(\theta)$, so the interpretation of results is direct.

First, we can note that the distortion will be larger if the local community decides to favor users further (second term in the equations above). Equation (12) shows that the rent given to the private operator depends on costs for the less efficient utility, and hence on extracted and sold volumes, because the information asymmetry increases the cost of production. The regulator is going to distort the loss index upward in order to reduce the rents (Equation (11)). And Equation (10) shows that it must reduce the sold water volume when the operator is inefficient. This means that it is in the interest of the local community to allow inefficient operators to increase their losses. This is because such strategy reduces the gain of an efficient operator to mimic an inefficient one.

6. Empirical Application

6.1. Data Description

We have data on 48 water utilities located in the Bordeaux region, in the south-west of France, for the years 1995, 1996, 1997 and 1998. Total number of observations is 192. All utilities are privately operated, i.e., delegated to private companies.

The main source of data comes from financial and technical reports made available by the Gironde Local Administration for Agriculture and Forest (Gironde DDAF). These reports contain information on total operating and maintenance costs, water volumes produced, electricity and labor inputs, as well as technical data on networks. The other source of data is a mail survey we directed toward local communities and the ESG private company marketing services. From this additional source, data on input costs and network features were complemented.

Water utility variable cost (VC) is the sum of labor (L), electricity (E), materials and other expenses (M). The fixed costs (FC) are the expenses related to investments and renewal expenses. Y_c and Y_l are computed respectively as the volume sold to final customers (V_c) and the volume lost (V_l) in the network, divided by the number of metered customers (n) in the local community. The loss index (I) is computed as $Y_l/Leng$, where $Leng$ is the network length (in km).

Labor input is defined as total hours worked in the year, unit labor price (including employer contribution to social security and pension benefit programs) (w_L) is obtained by dividing wage expenses by total hours worked. Unit price of electricity (w_E) is defined as the ratio of total electric power expenditures divided by the total quantity of electricity used in various production and distribution stages. Material expenses consist of different categories of heterogenous costs, obtained by consolidating accounts as different as stocking, maintenance work, and subcontracting. Because of data limitations and the problem of heterogeneity of this input, we choose to construct a price index for input materials (w_M) as a unit cost per cubic meter delivered.[3] Morever, we denote P the unit price charged by the private operator for the water service and AP the total average price (including operator's

Table 1. Sample descriptive statistics.

Variable	Unit	Mean	Standard deviation	Minimum	Maximum
VC	FF	1,478,950	1,675,248	134,566	10,210,000
FC	FF	281,256	412,250	0	2,535,000
w_L	FF/Hour	194.33	10.12	148.00	214.91
w_E	FF/KWh	0.48	0.20	0.15	1.70
w_M	FF/m^3	1.26	0.47	0.26	2.67
n	Unit	3097	3249	257	17210
$Leng$	Km	158	163	9	889
V_e	m^3	584,399	721,701	38,824	4,278,094
V_c	m^3	426,805	514,497	31,755	3,177,604
Y_e	m^3/Customer	181.10	33.05	107.92	305.08
Y_c	m^3/Customer	133.81	22.92	79.22	202.29
I	m^3/(Customer × Km)	0.66	1.03	0.07	8.16
P	FF/m^3	2.33	1.04	0.10	6.40
AP	FF/m^3	18.10	4.95	5.61	31.30

Notes: number of observations is 192 (48 utilities, 1995–1998).
VC and FC are respectively the variable costs and the fixed costs.
w_L, w_E and w_M denote unit cost of labor, energy, material and other expenses respectively.
n is total number of metered customers and *Leng* the water network length of water service.
V_e is volume extracted (and di stributed), V_c is water sold to customers, Y_e and Y_c are respectively these volumes divided by the number of customers.
I is the water network loss index, computed as $(Y_l)/Leng$.
P is unit water price paid to private operator and AP is total average price of water.
KWh is Kilo-Watt per hour.

fixed charge, municipality's price share, wastewater treatment charges and taxes) paid by the customer. Descriptive statistics for variables used are given in Table 1.

6.2. Empirical Results

The purpose of the empirical analysis is to provide consistent parameter estimates for technology (cost function), demand, and the distribution of the private information parameters θ. Based on these parameters, the first-order conditions corresponding to the optimal contract solutions can be simulated, given reasonable values for parameters π and the marginal damage functions with respect to V_c and I, respectively D_{V_c} and D_I. It is important to note that π and the parameters of D are the only information that cannot be retrieved from estimation of the cost and demand functions.

6.2.1. *Demand Estimation*

The first equation in the system of first-order conditions contains the inverse demand function, that must equal a given function of the marginal cost with respect to V_c plus the marginal damage with respect to V_c. To estimate such a demand function, it must be remembered that the inverse demand depends on the average price for water, whereas in our model, only the marginal price charged by the private operator is considered. For this reason, we assume a linear form for the inverse demand function.

Let AP denote average price, i.e., $AP = P(Y_c) + \tilde{P}$, where $\tilde{P}$ is the component of the average price corresponding to the operator's fixed charge, the local community share, wastewater treatment charge and taxes. We then have for community i,

$$AP_i = \alpha_i + \beta Y_{c,i} + v_i \Leftrightarrow P(Y_{c,i}) = \tilde{\alpha}_i + \beta Y_{c,i} + v_i,$$

where $\tilde{\alpha}_i = \alpha_i - \tilde{P}_i$. Hence, estimating parameters α_i and β from the genuine demand function allows to recover the relevant expression of the demand in terms of firm's price $P(Y_c)$.

We estimate the unit demand function above with our data on average price and average water demand per head, using the Fixed Effets procedure. The slope parameter β is estimated at -0.0567 and is significant (t-statistic of -5.18). We also perform the inverse regression of Y_c on average price. The slope parameter is estimated at -2.7845, with the same standard error as above. This corresponds to a price-elasticity of -0.347 at the sample mean, a value (for industrial and residential demands) a little higher than what is found in the empirical literature on residential water demand (see Nauges and Reynaud, 2001 or Nauges and Thomas, 2000) and smaller than what is found on industrial water demand by Renzetti (1992a). The effects $\tilde{\alpha}_i$, have an estimated mean of about 10.0436.

6.2.2. *Cost Estimation*

The second step in our empirical analysis consists in estimating the variable cost function for the water utilities. We choose a Translog functional form that allows enough flexibility in terms of substitutability between Y_c and Y_l (or the loss index I). The cost function is estimated within a system of equations including the cost shares for energy, and material and other expenses. We impose homogeneity by dividing cost and input prices w_E and w_M by the unit price for labor, w_L.

The question of endogeneity bias is particularly important to consider with our model when estimating the cost function, for the following reason. The observed cost should correspond to the optimal cost level for each utility, in the sense that it depends on optimal levels of control variables Y_c and I. But as the latter are defined as the solutions to first-order conditions (10) and (11), they implicitely depend on

θ and input prices if one solves for reduced-form equations. The hypothetical cost function reads:

$$C(\theta, V_c, I, w) = e^{\delta\theta} \times H[V_c(\theta, w), I(\theta, w), w], \tag{13}$$

where w is the vector of (normalized) input prices, and H the standard Translog functional form for cost. In logarithmic form for water utility i at period t, we have:

$$\log C_{it} = \delta\theta_i + \log H[V_c(\theta_i, w_{it}), I(\theta_i, w_{it}), w_{it}]. \tag{14}$$

Adding the conventional (i.i.d.) error term ε_{it} and incorporating the homogeneity restriction, we obtain

$$\begin{aligned}
\log\left(\frac{C_{it}}{w_{L,it}}\right) = {} & \beta_{V_c}\log V_{c,it} + \beta_I \log I_{it} + \frac{1}{2}\beta_{V_cV_c}(\log V_{c,it})^2 \\
& + \frac{1}{2}\beta_{II}(\log I_{it})^2 + \beta_{V_cI}\log I_{it}\log V_{c,it} + \beta_{w_E}\log\left(\frac{w_{E,it}}{w_{L,it}}\right) \\
& + \beta_{w_M}\log\left(\frac{w_{M,it}}{w_{L,it}}\right) + \frac{1}{2}\beta_{w_{EE}}\left(\log\left(\frac{w_{E,it}}{w_{L,it}}\right)\right)^2 \\
& + \frac{1}{2}\beta_{w_{MM}}\left(\log\left(\frac{w_{M,it}}{w_{L,it}}\right)\right)^2 + \beta_{w_{EM}}\log\left(\frac{w_{E,it}}{w_{L,it}}\right)\log\left(\frac{w_{M,it}}{w_{L,it}}\right) \\
& + \beta_{w_EV_c}\log\left(\frac{w_{E,it}}{w_{L,it}}\right)\log V_{c,it} + \beta_{w_EI}\log\left(\frac{w_{E,it}}{w_{L,it}}\right)\log I_{it} \\
& + \beta_{w_MV_c}\log\left(\frac{w_{M,it}}{w_{L,it}}\right)\log V_{c,it} \\
& + \beta_{w_MI}\log\left(\frac{w_{M,it}}{w_{L,it}}\right)\log I_{it} + \eta_i + \varepsilon_{it},
\end{aligned} \tag{15}$$

where η_i is the individual effect corresponding to $\delta\theta_i$. Because the private parameter θ is only defined up to a constant, we normalize δ to 1, so that $\eta_i = \theta_i$. It is clear from the equation above that variables $V_{c,it}$ and I_{it} are correlated with the individual effect η_i, as these three expressions depend on θ_i.

A convenient way to eliminate a possible endogeneity bias is to use the Within procedure for all equations in the system (including cost shares). To obtain a consistent variance-covariance matrix for parameters, we consider an Iterated SURE method jointly with the Within procedure. In other terms, the variance-covariance matrix of errors in the system is estimated iteratively, where errors refer to cost and cost share equations transformed with the Within procedure.

Estimation results are presented in Table 2. As a remark, we precise that we find estimates of marginal cost for each observation with respect to V_c which are all positive and with the two signs for the marginal cost with respect to I, as expected.

Table 2. Translog cost function estimation results.

Parameter	Estimated Coefficient		Standard Error	t-ratio
β_{V_c}	0.4020	(***)	0.0767	5.243
β_I	0.1946	(***)	0.0206	9.447
β_{W_E}	0.0754	(***)	0.0151	5.000
β_{W_M}	0.5659	(***)	0.0234	24.156
$\beta_{V_c V_c}$	–0.0654		0.0590	–1.107
β_{II}	0.0842	(***)	0.0123	6.866
$\beta_{V_c I}$	0.0666	(***)	0.0157	4.234
$\beta_{W_E W_E}$	0.0466	(***)	0.0043	10.957
$\beta_{W_M W_M}$	0.1970	(***)	0.0089	22.241
$\beta_{W_E W_M}$	–0.0403	(***)	0.0043	–9.304
$\beta_{V_c W_E}$	–0.0174		0.0125	–1.393
$\beta_{V_c W_M}$	–0.0617	(***)	0.0162	3.809
$\beta_{I W_E}$	–0.0073	(**)	0.0036	–2.014
$\beta_{I W_M}$	–0.0303	(***)	0.0060	5.067

Notes: number of observations is 192. Dependent variable: $\log(VC)$.
Homogeneity is imposed by dividing variable costs and input prices by w_L (labor price). $\bar{R}^2 = 0.9943$ (cost function). (**) and (***) denote respectively a parameter significant at 5 and 1%.

From these parameter estimates, estimated fixed effects $\hat{\eta}_i$ can be computed as $\overline{\log C_i} - \overline{\widehat{\log C_i}}$, where $\overline{\log C_i}$ denotes the individual average (across time periods) of $\log C_i$. These estimates are distributed between 7.6063 and 10.8748 on the sample, with a mean of 8.8734 and a standard deviation of 0.6798. Graphical inspection of the empirical θ plots reveals that a log-normal distribution is a sensible assumption to make, because of the right-skewed shape of the empirical density. Therefore, estimation results also allow to specify the complete distribution for the private information parameter, $F(\theta)$.

6.2.3. *Simulation of Optimal Contracts*

As noted above, estimated parameters for the demand and cost functions are not sufficient to derive solutions Y_c and I from the optimal contract equations. Two pieces of information are lacking, namely the weight π and the parameters of the damage function D. Once these parameters are set, the system of first-order conditions can be solved for Y_e and Y_c, recalling that I appearing in this system is equal to $Y_l/Leng$ and that $Y_e = Y_c + Y_l$.

Our economic model contains variables such as input prices (in the cost and cost share equations) and technical components (number of customers and network length), whose variation is not likely to be of interest when doing comparative statics. Consequently, we set input prices w_E, w_L and w_M, and network length *Leng* to their sample means. The number of customers n needs not be replaced by its sample mean, as n appears only in the definition of the water volumes per user.

It is difficult to decide what damage function one should select; a possible guideline is to choose a function for $D(.)$ that yields marginal damages for V_c and for unit loss I, that are fractions of the unit price for water. The approach used in this chapter is to select values for the damage D that lead to solutions (Y_e, Y_c) whose means are closest to their empirical counterparts.

We select values for θ using a grid ranging from the minimum of θ estimates with a stepwise of 0.2, and a maximum of 9.4.[4] For each θ, the term $F(\theta)/f(\theta)$ is computed from the log-normal distribution with mean 8.8734 and standard deviation 0.6798. We then compute solutions (Y_e, Y_c) for the average water utility by numerically solving the system of first-order conditions:

$$P(Y_C) - \frac{\partial C(\theta, V_c(\theta), I(\theta))}{\partial V_c} - \frac{\partial D(V_c(\theta), I(\theta))}{\partial V_c} - \left(\frac{2\pi - 1}{\pi}\frac{F(\theta)}{f(\theta)}\right)\frac{\partial^2 C(\theta, V_c(\theta), I(\theta))}{\partial V_c \partial \theta} = 0, \tag{16}$$

$$D_I + \frac{\partial C(\theta, V_c(\theta), I(\theta))}{\partial I} + \left(\frac{2\pi - 1}{\pi}\frac{F(\theta)}{f(\theta)}\right)\frac{\partial^2 C(\theta, V_c(\theta), I(\theta))}{\partial I \partial \theta} = 0. \tag{17}$$

The cost function parameters are replaced by their estimates from the Translog estimation procedure above.

For the damage function, we choose a quadratic form allowing to satisfy the convexity assumptions made above:

$$D(V_c, I) = b_0 + b_{V_c} V_c + b_I I + 0.5 b_{V_c V_c} V_c^2 + 0.5 b_{II} I^2 + b_{V_c I} V_c \times I,$$

so that the marginal damage functions are the following:

$$D_{V_c} = b_{V_c} + b_{V_c V_c} V_c + b_{V_c I} I,$$

$$D_I = b_I + b_{II} I + b_{V_c I} V_c.$$

We implement simulation so that the solutions (Y_e, Y_c) are closest to their empirical means that are equal respectively about 181 and 134, and choose coefficients so that marginal damages are equal to: $D_{V_c} = 0.1 + 3 \times 10^{-7} V_c + 2 \times 10^{-7} I$ and $D_I = 0.1 + 0.1 I + \times 10^{-7} V_c$.

To analyse the sensitivity of our results to various parameters, we compute optimal contract solutions with π ranging from 0.5 to 1.0. For $\pi = 0.5$, we have the First-Best (perfect information) solution.

Table 3. Optimal contract simulation results without damage

π	Y_e	Y_c	Rate of return (%)	Marginal Cost
0.50	194.92	145.14	74.43	1.28
0.60	192.69	142.00	73.65	1.31
0.70	191.05	139.64	73.03	1.33
0.80	189.77	137.78	72.52	1.35
0.90	188.76	136.27	72.09	1.36
1.00	187.94	135.02	71.73	1.37

Table 4. Optimal contract simulation results with damage.

π	Y_e	Y_c	Rate of return (%)	Marginal Cost	D_{Y_c}	D_I
0.50	186.75	140.85	75.40	1.30	0.231	0.042
0.60	184.85	137.68	74.44	1.33	0.228	0.045
0.70	183.35	135.29	73.72	1.35	0.226	0.047
0.80	182.16	133.40	73.15	1.37	0.224	0.048
0.90	181.18	131.87	72.67	1.38	0.223	0.050
1.00	180.37	130.59	72.27	1.40	0.221	0.051

Notes: θ is Log-normal with mean 8.8734 and standard deviation 0.6798.
Range for θ: [7.61; 9.41].
Volumes distributed and sold are in cubic meter per head.
Rate of return is defined as $100 \times Y_c/Y_e$.
Marginal damage corresponds to volume of cubic meter lost through the water network. Damage function is:
$D(V_c, I) = b_0 + 0.1V_c + 0.1I + 3 \times 10^{-7} V_c^2 + 0.1I^2 - 2 \times 10^{-7} V_c \times I$.
We compute the marginal cost of consumed water (V_c).
All figures except π are averages.

Tables 3 and 4 present the simulated values for Y_e and Y_c, the distributed and sold volumes of water per user (in cubic meter) respectively with and without damage. Several comments can be made. It can be seen that the level of consumed (and extracted) volume and the rate of return are decreasing with π as expected in the theorical framework above. The marginal cost of V_c increases with the distorsion due to asymmetric information. The introduction of a damage naturally leads to a decrease in the water volumes produced and a better network rate of return, but with an increasing marginal cost.

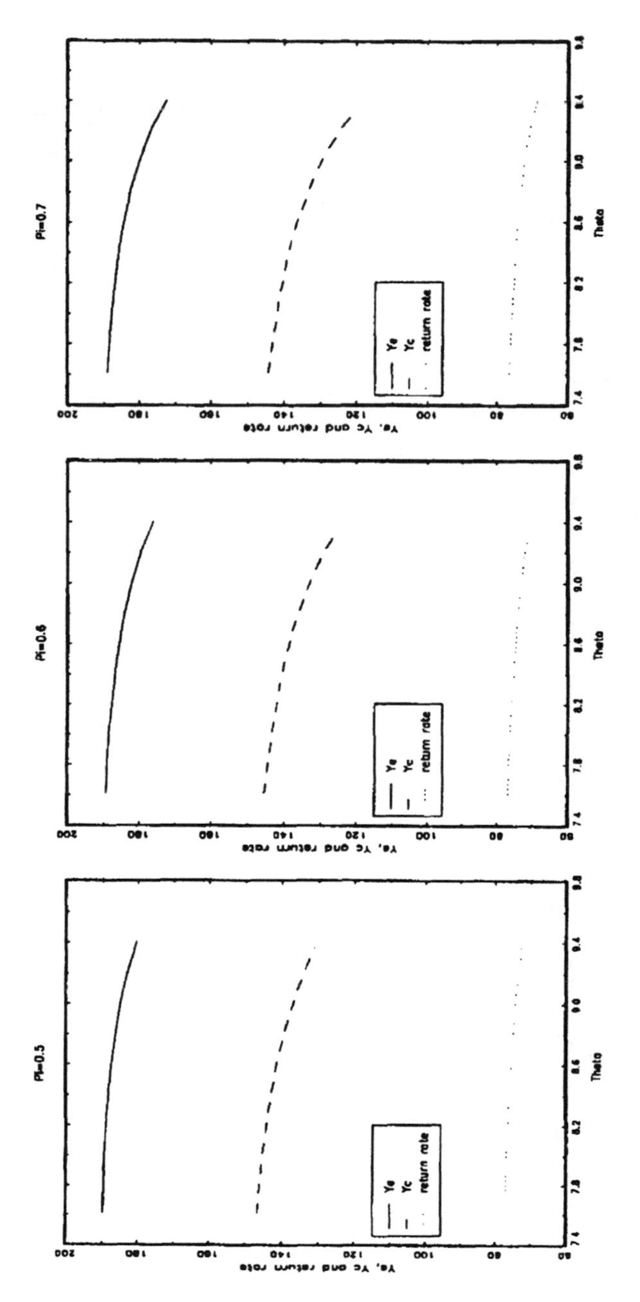

Figure 1. Optimal extracted volume, consumed volume and network rate of return.

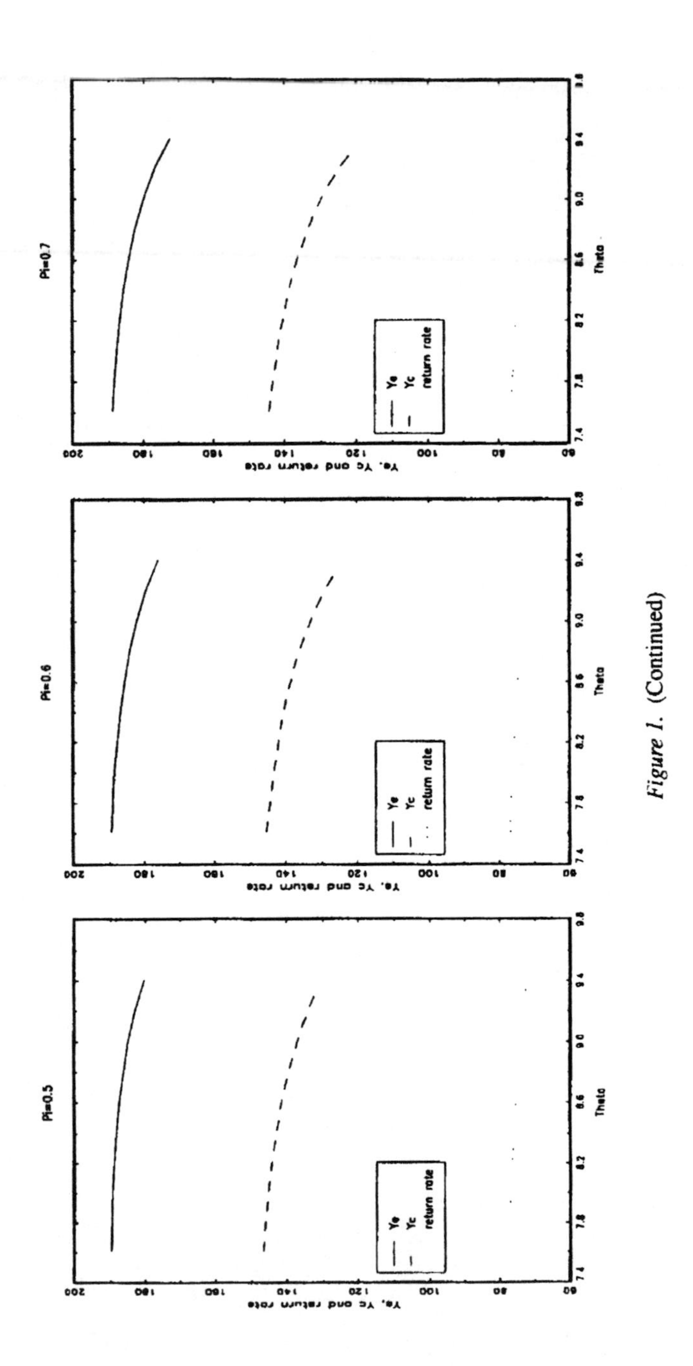

Figure 1. (Continued)

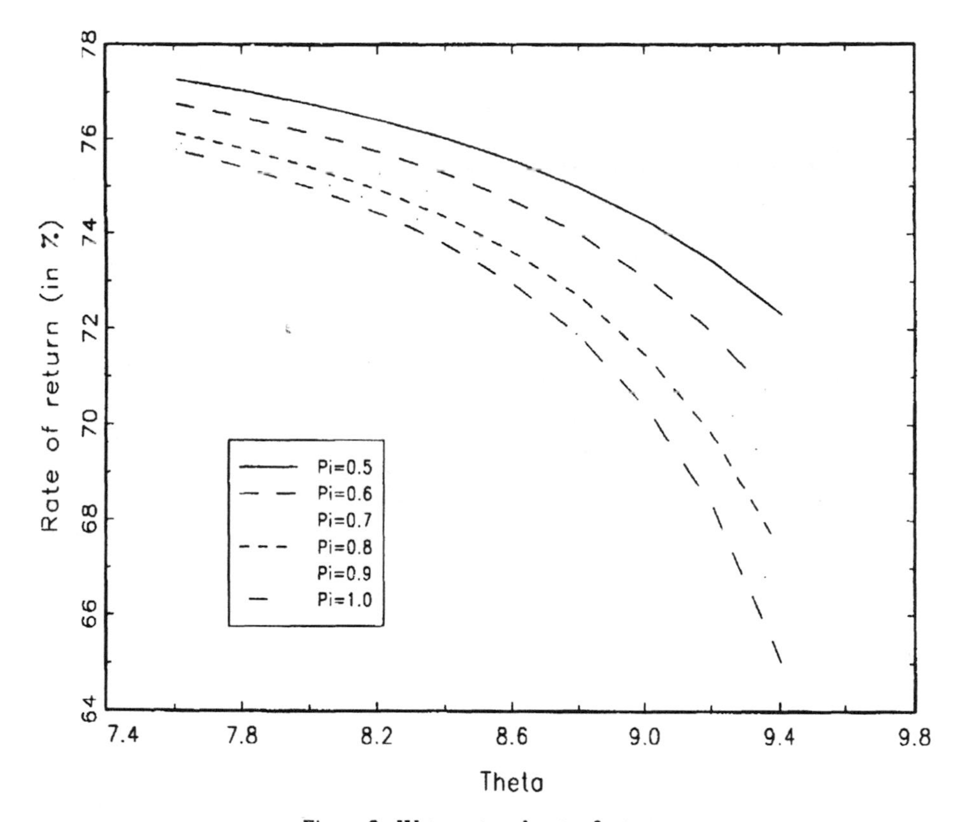

Figure 2. Water network rate of return.

Figure 1 presents the level of extracted water volume and sold volume per user, and the rate of return of the water network with respect to θ. Extracted and consumed volumes decrease with the private information parameter, and the rate of return decreases with θ. Hence, the optimal contract solution requires that the less efficient water utilities have a lower rate of return.

As π increases, the network rate of return diminishes. Figure 2 presents the average rate of return of the water network, for θ ranging from an estimated minimum ($\sim$7.6) and an upper bound set at 9.4. We can see that the rate of return is highest in the perfect-information case when $\pi = 0.5$, compared to asymmetric information cases when $\pi \neq 0.5$. This is due to the fact that the local community has to decrease the cost associated with the information rent by increasing the required network loss (or, equivalently, decreasing the rate of the return).

It can be seen from these simulation results that the presence of asymmetric information has an important impact on the contract solutions. The fact that the information rent to the private operator is increasing in the rate of return of the water

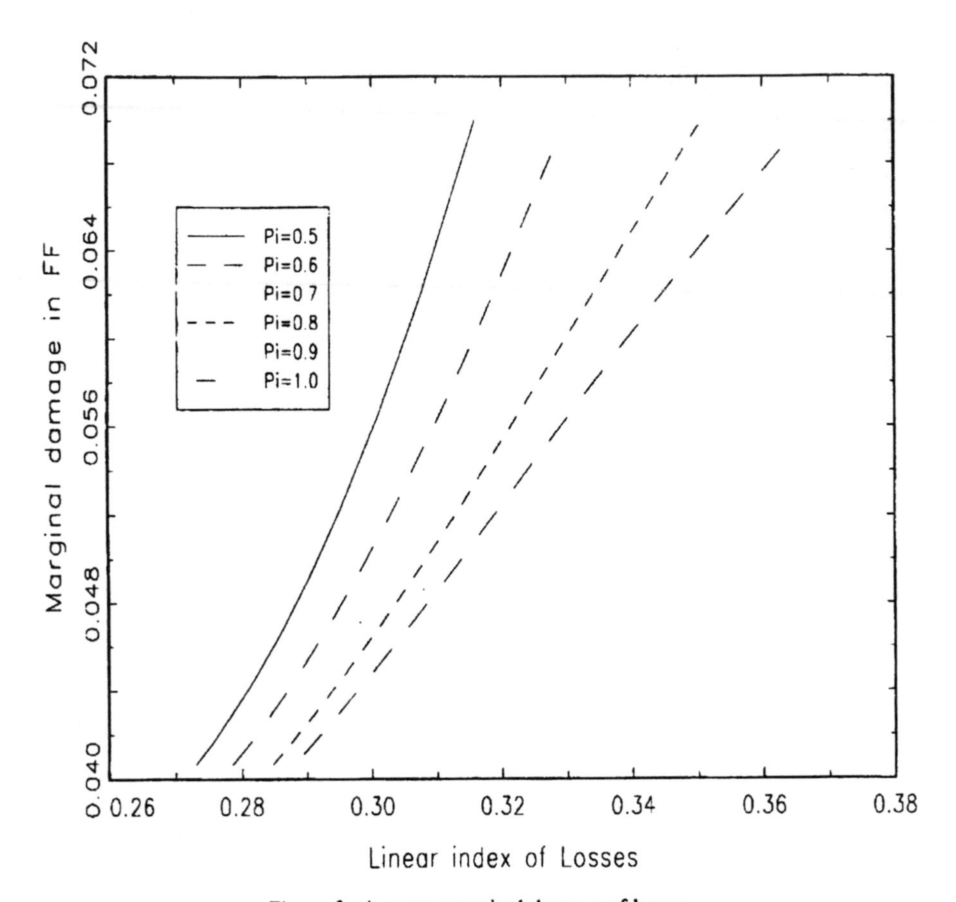

Figure 3. Average marginal damage of losses.

network has an important consequence for the local community. It is constrained to allow higher losses to reduce informational rents.

Figures 3 and 4 depict the average marginal damage for the same θ's, with different values for parameter π. If we consider an average consumption per user of 134 m^3, we have a value for marginal damage with respect to consumed water ranging roughly from 0.205 to 0.225 FF. Note that the marginal damage with respect to losses is decreasing with π. This result is fully consistent with the fact that the regulator has to provide the operator with incentives to keep losses at a high level: the stronger effect of asymmetric information, the less the regulator integrates the damage related to the losses. This value can be compared with the value of marginal cost of consumed water which is estimated around 1.35 to 1.40 FF on average.

Moreover, in order to account for damage to the community, the rate of return should be sufficiently high. This means that, for a given water output sold to customers, the distributed water level would need not to be too high. Hence, the larger

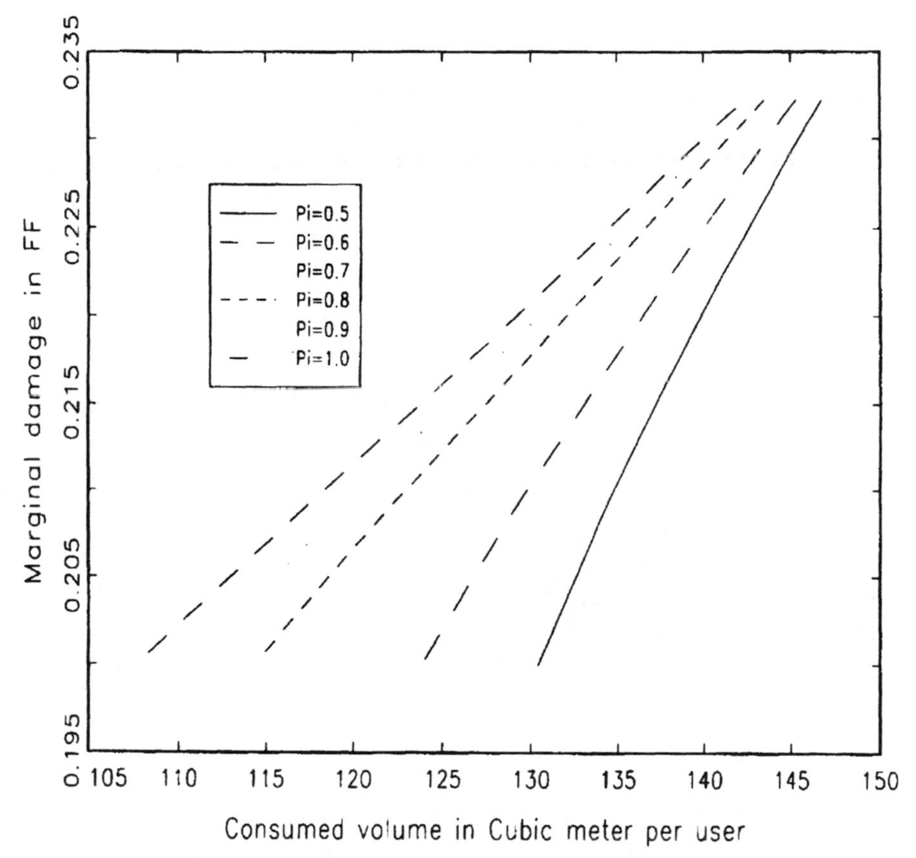

Figure 4. Average marginal damage of water consumption.

the private information parameter θ, the lower the rate of return. Also, since the information rent is increasing in π (because $(\partial(2\pi - 1)/\pi)/(\partial\pi) > 0)$, the rate of return in the water network is decreasing in the weight put to consumers (π).

7. Conclusion

In this chapter, we have presented an economic model of water supply regulation under asymmetric information between a principal (the local community) and an agent (the private operator). Moreover, we give a particular attention to the issue of network water losses. First, we have estimated cost and demand parameters, and the private information parameter distribution. Second, these estimates have allowed us to simulate the optimal contract solutions.

Our results can be summarized as follows. Consumed volume and the network rate of return decrease with the private information parameter. This means that the optimal contract solution requires that the principal allows the agent to increase its losses. The distortion due to asymmetric information is larger if the local community decides to favor users more than the private operator (increasing π). This is because the information rent to the private operator is decreasing when water losses increase. In the introduction, we give intuitions about the high level of losses observed in the water networks. A possible explanation is given by the relatively costly leak repairs compared to the ease of extracting more water. But we show that there exists another reason, that is, the principal requires the operator not to reduce losses.

Several extensions can be considered, in particular for the empirical part. It would be interesting to test for the presence of the private information in our sample. From the estimates of cost parameters, we could compute the predicted marginal costs that appear in the first-order conditions. The parameter π and the parameters of the damage function would then be the unknowns to estimate. A better methodology would be to estimate the full model in a single step. However, because we only know the distribution of θ and not individual values of the θ's, a simulated-based technique would be necessary, see McFadden and Ruud (1994).

Finally, we could also study in more detail the variation of customer surplus as well as the variation in private operator's rent.

Notes

[1] The operator can purchase an additional water volume to another water utility and also export water to another utility. However, because these volumes are negligible and that is not the major interest of our chapter, we do not explain the choice of the operator in purchasing and selling wholesale water to other utilities, so that the distributed volume (V_d) and the extracted volume V_e are equal.

[2] We take into account the fact that the firm chooses the effort e in order to minimize its costs.

[3] Price of material is simply defined as total expenses of different miscellaneous inputs such as subcontracting, stocking and maintenance work, divided by the distributed water volume.

[4] We do not use the whole interval of θ, because for highest values of θ the numerical solutions of the system do not converge.

References

Auriol, E., Ivaldi, M. and Kim, H. (1999) Advertising and regulation in the US local telephone market, Mimeo, Université des Sciences Sociales, Toulouse.

Baron, D.P. and Myerson, R.B. (1982) Regulating a monopolist with unknown costs, *Econometrica* **50**(4), 911–930.

Crain, W.M. and Zardkoohi, A. (1978) A test of the property-rights theory of the firm: Water utilities in the United States, *Journal of Law and Economics* **21**(2), 395–408.

Dalen, D.M. and Gomez-Lobo, A. (1996) Regulation and incentive contracts: An empirical investigation of the Norwegian vus transport industry, Working Paper No. W96/8, Institute for Fiscal Studies, London.

Dalen, D.M. and Gomez-Lobo, A. (1997) Estimating cost functions in regulated industries characterized by asymmetric information, *European Economic Review* **41**, 935–942.

Gagnepain, P. and Ivaldi, M. (1999) Incentive regulatory policies: The case of public transit systems in France, Mimeo, Université des Sciences Sociales, Toulouse.

Garcia, S. and Thomas, A. (2001) The structure of municipal water supply costs: Application to a panel French local communities, *Journal of Productivity Analysis* **16**(1), 5–29.

Gironde DDAF (1995, 1996, 1997, 1998) Technical and financial reports on private water utilities in Gironde, Internal Documents.

Hayes, K. (1987) Cost structure of the water utility industry, *Applied Economics* **19**, 417–425.

Kim, H.Y. (1987) Economies of scale in multiproduct firms: An empirical analysis, *Economica* **54**(214), 185–206.

Laffont, J.J. and Tirole, J. (1986) Using cost observation to regulate firms, *Journal of Political Economy* **94**(3), 614–641.

Lau, L.J. (1976) A characterisation of the normalized restricted profit function, *Journal of Economic Theory* **12**, 131–163.

Lavergne, P. and Thomas, A. (1997) Semiparametric estimation and testing in models of adverse selection, with an application to environmental regulation, Mimeo, Université des Sciences Sociales, Toulouse.

Mann, P.C. and Mikesell, J.L. (1976) Ownership and water system operation, *Water Resources Bulletin* **12**(5), 995–1004.

McFadden, D. and Ruud, P.A. (1994) Estimation by simulation, *Review of Economics and Statistics* **76**(4), 591–608.

Nauges, C. and Reynaud, A. (2001) Estimation de la demande domestique d'eau potable en France, *Revue Economique* **52**(1), 167–185.

Nauges, C. and Thomas, A. (2000) Privately-operated water utilities, municipal price negotiation, and estimation of residential water demand: The case of France, *Land Economics* **76**(1), 68–85.

Renzetti, S. (1992a) Estimating the structure of industrial water demands: The case of Canadian manufacturing, *Land Economics* **68**(4), 396–404.

Renzetti, S.J. (1992b) Evaluating the welfare effects of reforming municipal water prices, *Journal of Environmental Economics and Management* **22**, 147–163.

Thomas, A. (1995) Regulating pollution under asymmetric information: The case of industrial wastewater treatment, *Journal of Environmental Economics and Management* **28**, 357–373.

Wolak, F.A. (1994) An econometric analysis of the asymmetric information, regulator-utility interaction *Annales d'Economie et de Statistiques* **34**, 13–69.

4.3. *Water Policy and Water Practice – Second Best*

Policy Instruments for Groundwater Management for Agriculture and Nature in the Netherlands

PETRA J.G.J. HELLEGERS and EKKO C. VAN IERLAND

1. Introduction

In the Netherlands nature reserves are suffering from desiccation, as a result of falling groundwater levels. Intensification of agricultural activities contributed to the lowering of groundwater levels by intensified drainage of adjacent farmland to raise production and by increased agricultural groundwater extraction for consumption and sprinkling. In the presence of growing conflicts between agriculture and nature with respect to groundwater management there is a pressing need to understand the potential role of policy instruments for groundwater management. Insight into the suitability of policy instruments to achieve optimal groundwater level and extraction management in the Netherlands is, however, missing.

Economic literature on the role of policy instruments for groundwater management in the Netherlands has been limited, especially with respect to agriculture. Wiersma (1998) investigated a more efficient groundwater allocation procedure based on regional auctions for Dutch groundwater extraction permits. He does not consider agricultural extractions in his analysis, because they are only temporary and will be reduced by sprinkling bans. There is, however, extensive literature available (e.g. Michelson and Young, 1993; Dinar and Wolf, 1994; Strosser, 1997) on the role of groundwater allocation mechanisms in other countries.

Since policy instruments can 'queue' users and restrict the transfer and trading of water and adoption of modern technologies, it is important to study its current and potential role. *The aim of this chapter is therefore to test the suitability of various instruments for groundwater level and extraction management in the Netherlands.* We will study the suitability of three economic instruments, one regulatory instrument, one suasive instrument and a change in the institutional environment. To provide insight into the suitability of these policy instruments, they will be reviewed against six performance criteria: environmental effectiveness, economic efficiency, technical efficiency, administrative feasibility, equity and acceptability.

The structure of this chapter is as follows. In Section 2 current instruments for groundwater level management are discussed and the suitability of alternative instruments is tested against the six policy review criteria. In Section 3 current

P. Pashardes et al. (eds.),
Current Issues in the Economics of Water Resource Management, 221–232.

instruments for groundwater extraction management are discussed and the suitability of alternative instruments is tested in the same way. Section 4 contains the discussion and conclusions.

2. Suitability of Policy Instruments for Groundwater Level Management

In this section we will test the suitability of instruments for groundwater level management. It is important to note that economic instruments are at present not used for groundwater level management in the Netherlands. We will first explain how current instruments work. Then we will discuss whether improvements are desirable and we will propose some alternatives. The performance of each of the instruments will be discussed separately. First we describe the prerequisites for instruments to be effective. After that we discuss whether it improves the economic and technical efficiency. Next we discuss the administrative feasibility. We also discuss whether the financial impact is equitably distributed among affected parties, which affects the acceptability. Finally, practical implementation problems will be addressed. The results of these tests are summarised in Table 1.

Table 1. Suitability of instruments for groundwater level management[1].

Instrument	Effectiveness	Economic efficiency	Technical efficiency	Administrative feasibility	Equity	Acceptability
1A Tax	+/–	+	+	–	+/–	–
1B Tradable rights	+	+	+	–	+/–	+/–
1C Subsidy	+/–	–	–	–	–	+
1D Standard	+	–	–	–	+/–	–
1E Agreements	+/–	+	+	+	+	+/–
1F Responsibilities	+	+	+	+	+	+

[1]Three categories are distinguished: '+' stands for a positive impact, '–' for a negative impact and '+/–' indicates the possibility of a positive as well as a negative impact. A '+' can be interpreted as follows:
'+' for environmental effectiveness means that the instrument reaches its environmental objective.
'+' for economic efficiency indicates that the value of the marginal product of water is equal among users.
'+' for technical efficiency means that the instrument provides incentives to adopt modern technologies.
'+' for feasibility means that the instrument is easy to implement, monitor and enforce against low cost.
'+' for equity means that the costs and benefits are equitably distributed among affected parties.
'+' for acceptability means that affected parties accept the instrument without serious resistance.

2.1. Economic Instruments

(1A) Theoretically *a tax* can be used to restrain farmers from lowering the groundwater level below a certain standard. The effectiveness of a tax depends on the right estimation of the marginal tax level and on how risk averse farmers are with respect to wet damage. A differentiated tax level has to be created, because of local differences in both the monetary value of reserves and vulnerability of the environment to changes in the groundwater level. An advantage of a tax is that it improves both the economic and the technical efficiency. Administrative cost will be high, since a differentiated tax is not easy to control and monitor. The financial impact on affected parties depends on the restitution of revenues, which affects the acceptability. Finally, there are practical implementation problems. It is hard to define a good basis for a tax. A volumetric tax on drainage is complicated, since it is not easy to measure the amount of drainage water. A tax on a change in the groundwater level is also complicated, because external factors also affect the groundwater level. Besides farmers do not directly pay a price for *in situ* services of groundwater[1] and usually groups of farmers will be affected by changes in the groundwater level. Charging water-boards for lowering surface water levels will not influence an individual farmer's behaviour, but it will affect the strategy of groups of farmers represented in the governing body of water-boards.

(1B) *Tradable rights* for lowering the groundwater level are ceilings on lowering that, once initially allocated, can be traded subject to a set of prescribed rules. The environmental objective is the starting point. Tradable rights improve the economic and technical efficiency, since the market determines the price of the right in a dynamic way (Pearce and Turner, 1990). The high demand for administrative institutions is a major disadvantage. The financial impact on affected parties depends on the initial allocation of rights, which is also determining for the acceptability. The use of tradable rights for lowering the groundwater seems to be complicated in practice, since the impact of changes in the groundwater level on agricultural production and nature depends on location specific circumstances. To avoid the transfer of rights to areas sensitive to desiccation, trading among areas has to be restricted. Here we face a dilemma: on the one hand the market approach is embraced, but on the other hand we need a trade institution for guided trading (Kruitwagen et al., 2000).

(1C) *A subsidy* is a reward for meeting a certain groundwater level, which is higher than the desired standard. Subsidies are not economically efficient, they are disturbing and do not provide incentives for the adoption of modern technologies. The acceptability is no problem, since participation in subsidy schemes is voluntary and because of the financial implications. Implementation problems are similar to those of a tax.

Farmers currently receive payments for drastic income losses due to higher groundwater levels. These payments are called subsidies, which is a misleading term in this respect, since these payments do not provide incentives to lower the groundwater level, but aim to balance financial implications on affected parties. Whether this is justified, depends on the allocation of rights to lower groundwater levels. Since farmers currently receive financial compensation for private benefits foregone, we implicitly ascribe rights to lower groundwater levels to agriculture. These payments avoid that farmers leave the region due to income losses. In that respect it provides incentives to affect individual farmer's behaviour.

2.2. Regulatory Instruments

(1D) A legal groundwater level standard can be introduced. It will be effective, if farmers face substantial monetary penalties for lowering the groundwater level below this standard. The standard does not improve the economic efficiency and does not provide incentives to innovate. The financial impact is not always equitably distributed among affected parties, since there are differences in the vulnerability of areas to changes in drainage. Differentiated standards will pose a large burden on the administrative capacity. Usually serious resistance is raised against the introduction of standards.

2.3. Suasive Instruments

(1E) Currently voluntary agreements about drainage control are established between farmers and governmental organisations, representing the interests of nature. Participation in such drainage control programmes is encouraged by means of positive incentives (a restitution of taxes). Under such programmes education activities were started to convince farmers of the advantages of fine-tuned drainage control. Less drainage leads for instance to a better utilisation of minerals by the crop, which encourages the adoption of sophisticated drainage management tools, like adjustable weirs. Such voluntary agreements about drainage control seem to very be suitable and efficient, since they use specialised knowledge of participants about local conditions. When costs and benefits are not equitably distributed among affected parties, both parties can bargain about compensation payments. The allocation of such payments depends on the assignment of rights. The acceptability is no problem, since it is a voluntary regime. Because of all these advantages, voluntary agreement initiatives for drainage control could be highly recommended. Participation of farmers in planning and decision-making at the local level is becoming more common. The principle of allowing the individual members of agricultural organisation and water-boards to make decisions on issues that affect them rather than leaving those decisions to be made by the whole group, the so-called principle of subsidiarity, is widely accepted.

2.4. Change in the Institutional Environment

(1F) Until 1992, agricultural interests have mainly dominated within the governing body of the water-boards as a result of the strong financial interests of this sector. Since 1992 environmental and nature organisations are better represented in the governing body of the water-boards. They represent the interests of nature with respect to groundwater level management. Interests of nature have often been ignored in the decisions about surface water levels, which directly effect groundwater levels. If water-boards become legally responsible for groundwater level management and if more interests are represented within the water-boards, various interests with respect to groundwater levels will be better balanced and can become guiding for decisions about surface water levels. This might improve the economic efficiency in an equitable manner. Administrative costs of such a change seem to be low, although changes in the institutional environment are generally not so easy to implement. The acceptability seems no problem as long as the basic principle of water-boards 'interest, payment and authority' is remained.

3. Suitability of Policy Instruments for Groundwater Extraction Management

In this section the suitability of policy instruments for groundwater extraction management will be tested in the same way as instruments for groundwater level management in Section 2. Table 2 shows the test results of instruments for groundwater extraction management.

3.1. Economic Instruments

(2A) Extraction is currently subject to two acts for a financial contribution to the government.

Table 2. Suitability of instruments for groundwater extraction management[1].

Instrument	Effectiveness	Economic efficiency	Technical efficiency	Administrative feasibility	Equity	Acceptability
2A Tax	+/–	+	+	–	+/–	–
2B Tradable rights	+	+	+	–	+/–	+/–
2C Subsidy	+/–	–	–	–	–	+
2D Ban	+	–	–	–	+/–	–
2E Agreements	+/–	+	+	+	+	+
2F Change rights	+	+	+	+/–	–	–

[1]See Table 1 for the explanation of the '+' for the various performance criteria.

- Since 1995, extraction is subject to a tax, under the 'Act Taxes on Environmental Basis' (Wet Belasting op Milieu-Grondslag), which has to be paid to the central government. It is not a regulating tax, since the main aim is not to reduce usage, but to generate funds. It works with two tariffs: EURO 0.15 per m^3 for waterworks and EURO 0.08 per m^3 for other extractors. There is an annual tax-free threshold of 40,000 m^3 for extractions that are used for sprinkling. Extractions with pumps with a capacity of less than 10 m^3 an hour and extractions which are infiltrated back to the resource are also exempted from the tax. As extractions of waterworks are not excepted from this tax, the price of tap water has increased. This has provided incentives to farmers to sink their own wells, which means that such a tax is not very effective. Such diffuse extractions also have a negative impact on groundwater quality. We should therefore be careful with the creation of exceptions.
- Since 1985, extraction is subject to a levy, under the 'Groundwater Act' (Grondwater Wet), which has to be paid to provinces. The levy-free threshold and tariffs vary among provinces due to local differences and are subject to change. Tariffs are relatively small, for instance EURO 0.0136 per m^3 in the province of Noord-Brabant in 2001.

Currently only a small percentage of farmers (about 2%) exceed the tax-free threshold, and are subject to the tax (Van Staalduinen et al., 1996). The main part of agricultural extraction is also not subject to the levy under the Groundwater Act. Most farmers currently only pay the energy costs of lifting water from the stock to the field (i.e. EURO 0.04 per m^3). The current pricing system of water implies that externalities, which arise due to agricultural groundwater extraction, are not yet fully internalised in the price of water. It is not efficient, since farmers maximise individual (instead of social) current (instead of future) profit and pump water until its marginal net benefit is zero (in the absence of bans).

An alternative is to impose a higher volumetric effective tax, which can be considered as a kind of water pricing reform. Whether such a tax is justified depends on the allocation of property rights for extraction, whereas it is not clear who owns the property rights. The theoretical framework of the optimal tax level is clear (the Pigouvian tax upon the activities of the generator of an externality has to be equal to the marginal externality cost produced by that activity), but difficulties emerge if a proper tax level has to be determined in reality. A differentiated tax system has to be created, because of local differences in the availability and quality of groundwater and the role groundwater plays with respect to terrestrial ecosystems. The effectiveness depends on how risk avers farmers are with respect to sprinkling and on the price elasticity of water demand. This price elasticity seems to be too low to do an effective job, because sprinkling is even currently not always economically profitable. Other aspects, like peace of mind and risk attitude of farmers, also play a role in the decision to sprinkle.

Such a tax is easy to adjust and increases therefore flexibility. It reduces extraction where it is most efficient and improves the technical efficiency, since it raises the water price. The diffuse, irregular and smaller extractions by agriculture together with the differentiated tax system will pose a large burden on the administrative capacity. The financial impact of taxes depends on restitution of revenues. The acceptability depends on how influential parties are.

(2B) Currently the transfer of rights to extract groundwater is already possible in the province of Brabant. *Tradable rights* are a restricted number of agricultural extraction rights that, once initially allocated by authorities, can be traded subject to a set of prescribed rules. Transition to agricultural water markets while diverting water away from agricultural use, may decrease agricultural sector's well-being to some extent, but is desirable from a social point of view (Shah and Zilberman, 1992). Tradable rights improve the economic and technical efficiency, since the market determines the price of the right in a dynamic way. The high demand for administrative institutions is a disadvantage of water markets. An equitable introduction of water markets is, however, hard to establish. Rights can for instance be auctioned off, so that the authorities reap all the rent from new entitlements. An alternative is to allow senior rights owners to sell their water to buyers and benefit from the revenues of the sales. The financial impact on affected parties is determining for the acceptability. The use of tradable rights for groundwater extraction seems to be complicated in practice, since the impact of groundwater extraction on the desiccation of nature depends on location specific circumstances. To avoid the transfer of extraction rights to regions sensitive to desiccation, guided trading is required. As the market will not take differences in susceptibility of nature reserves to desiccation into account, it is necessary to intervene in the market to safeguard the environmental targets. The price of extraction rights is closely related to the heterogeneity criterion of water, geographic area, characteristics of the local market, the size of the transaction, the number and size of potential traders and the information and searching costs that are involved in the transaction (Colby et al., 1993).

(2C) *A subsidy* can be provided for meeting a certain volume of extraction, which is smaller than the desired standard. Subsidies fail to give a clear sign of real scarcity to farmers and provide no incentives for adoption of modern technologies. Administrative costs are high. The acceptability is no problem, since participation in subsidy schemes is usually voluntary and because of the financial implications.

3.2. Regulatory Instruments

(2D) Sprinkling bans currently aim to reduce low-value agricultural groundwater extraction. They divert water away from current agricultural use to non-agricultural and/or future use. These bans differ per province and vary with respect to the source of water used (groundwater *versus* surface water), crop type (grass *versus* arable), soil type (sandy *versus* clay) and time period (part of the year and day). Bans especially aim to reduce groundwater use for sprinkling of grass on sandy soils in areas sensitive to desiccation during periods of drought. Farmers are not compensated financially for income losses due to sprinkling bans, which means that the extraction rights are not implicitly ascribed to agriculture. Current bans are only rough restrictions. There are for instance no arable crop-specific sprinkling bans. Rough bans are suitable for a quick reduction in extraction and to ban a certain extractor.

An alternative is to fine-tuned bans to resource, region, soil, crop and time specific circumstances in such a way that they will allocate water efficiently. Bans provide incentives to change farming practices (like the cropping pattern), but it does not provide incentives to adopt modern technologies. The financial impact is not always equitably distributed among affected parties, since there are differences in the vulnerability of areas to extraction (and therefore in bans). Differentiated bans will pose a large burden on the administrative capacity. Usually serious resistance is raised against the introduction of bans.

3.3. Suasive Instruments

(2E) *Voluntary agreements* currently induce participation in irrigation scheduling programmes. A management tool for irrigation scheduling was developed and tested in 1995, often referred to as the sprinkling planner. Irrigation scheduling gives farmers a better insight into the moisture regime of their plots, the best timing to start irrigation and the best water dose and prevents over-irrigation and thus increases the irrigation effectiveness. It is not likely that farmers will adopt the sprinkling planner under the low water prices they currently face. Motives for not adopting it are the investment costs it entails, its complexity and the effort that its use will require. Practical test results showed that indeed only a small group of farmers would adopt the sprinkling planner of their own accord (Boland et al., 1996). Farmers are therefore subsidised and education and persuasion activities were started to induce participation (carrot approach). Persuasion is hard since farmers often behave myopic due to the competition in the sector. Besides, there will be no sprinkling bans, if a certain diffusion rate will be met (stick approach), i.e. if a certain number of farmers adopt the planner within a certain time period. Those who do not participate also benefit from the absence of bans. They can be considered as 'free riders'. Another very promising voluntary agreement we recommend is a commitment between farmers and nature conservationists on the extraction of groundwater. It is only effective if such agreements are really established. It reduces extraction in an efficient and equitable way against low administrative costs. The acceptability is no problem, since it is a voluntary compliance regime.

3.4. Change in the Institutional Environment

(2F) The current extraction rights system is based on free extraction permits granted by local authorities (provinces) in the past. These permits can be considered as historical extraction rights ('grand-fathering rules'). The system only refused an extraction right when the proposed extraction could damage other users (Perdok and Wessel, 1998), whereas damage of extraction to ecosystems was until very recently not taken into account. Nature is jeopardised under such a system. The current groundwater extraction rights system is not efficient, since current allocation rules are based on a 'queuing' system that restricts the trading of rights. It 'queues' in particular users of *in situ* services of groundwater, like nature.

Changes in the assignment of the extraction rights can reduce extraction in an efficient way in areas sensitive to desiccation. A restricted number of groundwater extraction rights have to be redistributed among farmers with the intention to allocate rights in an efficient and equitable way, which is not easy to establish. Differences in the vulnerability of ecosystems to extraction should be taken into account when extraction rights are assigned. Changes in rights are not so easy to modify and adjust. Besides, involved parties are generally very sensitive to changes in the rights system and it may therefore encounter serious resistance.

4. Discussion and Conclusions

4.1. Discussion

In this chapter groundwater policy instruments were discussed separately and it becomes clear that each instrument has its advantages and disadvantages. Policy instruments can, however, be combined in such a way that they reinforce each other. A mixture of policy instruments can be very fruitful (OECD, 1991). There are many cases where economic instruments are applied in conjunction with other instruments. Especially combinations with regulatory instruments are quite common. It is important to realise that economic instruments are only potentially effective in a limited number of cases. There will often be a role for non-market devices, working in tandem with economic incentives (FAO, 1994). Besides changes in the institutional environment might be needed. This makes it necessary to consider all economic, legal, institutional, political, social and cultural constraints. It is difficult to give guidelines for the use of instruments under certain circumstances. An empirical analysis of the region specific circumstances is necessary to provide insight into the optimal policy instrument mix.

Policy reform is usually conditional upon the size of social gains relative to transaction costs. Transaction costs diminish net-benefits of reforms. Transaction costs of water transfers are often high in absolute terms, but also in relative terms as compared to the low economic value of water per unit of volume due to the bulkiness characteristic of water (FAO, 1994). Zilberman et al. (1997) showed that the transition from a water queuing system to a tradable water rights system may lower social welfare, in cases of high transaction costs. This occurs when potential gains are lost to search and negotiation costs. Unfortunately, little empirical work has been done to determine the magnitude and form of transaction costs or what factors influence transaction costs. On the one hand, some fixed costs are likely to be associated with transactions (like legal costs and registration fees), and this would suggest declining marginal costs. On the other hand, search costs are more likely to be characterised by rising marginal costs as potential traders search first those traders who are most willing to trade (Renzetti, 2000). Because transaction costs are so important, the choice of institutional arrangements for dealing with these costs is critical.

4.2. Conclusions

In this chapter we studied the suitability of current and alternative policy instruments for groundwater level and extraction management. Unambiguous statements about the suitability of policy instruments are, however, hard to make without an empirical analysis, since the suitability differs locally considerably due to region specific circumstances. It depends on characteristics of the situation (WRR, 1992), like the relative power of target groups.

Nevertheless, from our theoretical analysis it becomes clear that economic instruments are very suitable for groundwater extraction management, since it is a demand-based problem, whereas the use of economic instruments for groundwater level management will be complicated in practice. Firstly, because it is hard to define a good basis for a tax. Secondly, it will be hard to influence the individual farmer's behaviour, since farmers do not directly pay a price for *in situ* services of groundwater and usually groups of farmers will be affected by changes in the groundwater level. Finally, groundwater level management problems are characterised by differences in vulnerability of areas to changes in drainage. In spite of this, economic instruments can affect the outcome of negotiations between nature organisations and agriculture about the groundwater level in the governing body of the water-boards.

Changes in the institutional environment and voluntary agreements seem to be more suitable for groundwater level management than economic instruments. It is recommended to make water-boards legally responsible for groundwater level management, so that the, often conflicting, interests with respect to groundwater level management will be better balanced. Voluntary agreements between groups of farmers and nature conservationists about drainage control seem also to be very promising, since it uses specialised knowledge of participants about local conditions and it can encourage the adoption of management tools, like adjustable weirs. It is important to know in this respect who owns the rights to lower groundwater levels. Since farmers are currently compensated financially for income losses due to higher groundwater levels, we implicitly ascribe rights to lower groundwater levels to agriculture.

The current historical groundwater extraction rights systems together with the low groundwater prices encourage low-value agricultural groundwater usage, whereas sprinkling bans and irrigation scheduling currently aim to reduce low-value use of groundwater. These extraction instruments are less efficient than a system that considers externalities in the price of water or diverts water away from agriculture while encouraging trading. Economic instruments seem therefore to be very promising for groundwater extraction management.

Note

[1] Besides extractive services, groundwater also provides *in situ* services, which occur as a consequence of groundwater remaining in place. For instance the capacity of groundwater to; prevent subsidence of the land, buffer against periodic water shortages, protect against seawater intrusion, protect water quality by maintaining the capacity to dilute, and facilitate habitat and ecological diversity.

References

Boland, D., Bleumink, H. and Buys, J.C. (1996) Naar een beregeningsplanner voor agrariërs II, Toetsing in the praktijk (Towards an sprinkling planner for farmers II, Testing it in practise), Centrum voor Landbouw en Milieu Report 235, Utrecht.

Colby, B.G., Crandall, K. and Bush, D.B. (1993) Water rights transactions: Market values and price dispersions, *Water Resources Research* **29**(6), 1565–1572.

Dinar, A. and Wolf, A. (1994) International markets for water and the potential for regional cooperation: Economic and political perspectives in the Western Middle East, *Economic Development and Cultural Change* **43**, 44–66.

Food and Agriculture Organisation (FAO) (1994) Reforming water resources policy – A guide to methods, processes and practices, FAO Irrigation and Drainage Paper No. 52, Rome.

Kruitwagen, S., Folmer, H., Hendrix, E., Hordijk, L. and van Ierland, E. (2000) Trading sulphur emissions in Europe: 'Guided bilateral trade', *Environmental and Resource Economics* **16**, 423–441.

Michelson, A.M. and Young, R.A. (1993) Optioning agricultural water rights for urban water supplies during drought, *American Journal of Agricultural Economics* **75**, 1010–1020.

Organisation for Economic Co-operation and Development (OECD) (1991) *Environmental Policy: How to Apply Economic Instruments*, OECD, Paris.

Pearce, D. and Turner, R. (1990) *Economics of Natural Resources and the Environment*, Harvester Wheatsheaf, Hertfordshire.

Renzetti, S. (2000) *An Empirical Perspective on Water Pricing Reforms*, in A. Dinar (ed.), *The Political Economy of Water Pricing Reforms*, Oxford University Press, New York, pp. 123–140.

Perdok, P.J. and Wessel, J. (1998) Netherlands, in F.N. Correia (ed.), *Institutions for Water Resources Management in Europe*, Balkema, Rotterdam, pp. 327–447.

Shah, F. and Zilberman, D. (1992) Queuing vs. markets, Department of Agricultural and Resource Economics, University of California.

Strosser, P. (1997) Analysing alternative policy instruments for the irrigation sector; An assessment of the potential for water market development in the Chishtian Sub-division, Ph.D Thesis, Wageningen University.

Van Staalduinen L.C., Hoogeveen, M.W., Ploeger, C. and Dijk, J. (1996) Heffing van grondwaterbelasting via een forfait; Een onderzoek naar de mogelijkheden voor de land- en tuinbouw, Publicatie 3.163, Agricultural Economics Research Institute, The Hague.

Wetenschappelijke Raad voor het Regeringsbeleid (WRR) (1992) Milieubeleid: Strategie, instrumenten en handhaafbaarheid, Rapporten aan de Regering, 41, SDU, Den Haag.

Wiersma, D. (1998) Towards a more sustainable use of groundwater by means of economic incentives. A case study of The Netherlands, in S. Dwyer, U. Ganslosser and M. O'Connor (eds.), *Life Science Dimensions; Ecolgical Economics and Sustainable Use*, Filander Verlag, Fürth, Germany.

Zilberman, D., Chakravorty, U. and Shah, F. (1997) Efficient management of water in agriculture, in D.D. Parker and Y. Tsur (eds.), *Decentralization and Co-ordination of Water Resource Management*, Kluwer Academic Publishers, Dordrecht, pp. 221–246.

Economy & Environment

1. F. Archibugi and P. Nijkamp (eds.): *Economy and Ecology: Towards Sustainable Development.* 1989 ISBN 0-7923-0477-2
2. J. Bojö, K.-G. Mäler and L. Unemo: *Environment and Development: An Economic Approach.* 1990 ISBN 0-7923-0802-6
3. J. B. Opschoor and D. W. Pearce (eds.): *Persistent Pollutants: Economics and Policy.* 1991 ISBN 0-7923-1168-X
4. D.J. Kraan and R. J. in 't Veld (eds.): *Environmental Protection: Public or Private Choice.* 1991 ISBN 0-7923-1333-X
5. J.J. Krabbe and W.J.M. Heijman (eds.): *National Income and Nature: Externalities, Growth and Steady State.* 1992 ISBN 0-7923-1529-4
6. J. Bojö, K.-G. Mäler and L. Unemo: *Environment and Development: An Economic Approach* (revised edition). 1992 ISBN 0-7923-1878-1
7. T. Sterner (ed.): *Economic Policies for Sustainable Development.* 1994 ISBN 0-7923-2680-6
8. L. Bergman and D.M. Pugh (eds.): *Environmental Toxicology, Economics and Institutions.* The Atrazine Case Study. 1994 ISBN 0-7923-2986-4
9. G. Klaassen and F.R. Førsund (eds.): *Economic Instruments for Air Pollution Control.* 1994 ISBN 0-7923-3151-6
10. K. Uno: *Environmental Options: Accounting for Sustainability.* 1995 ISBN 0-7923-3513-9
11. K. Uno and P. Bartelmus (eds.): *Environmental Accounting in Theory and Practice.* 1997 ISBN 0-7923-4559-2
12. J.C.J.M. van den Bergh, K.J. Button, P. Nijkamp and G.C. Pepping: *Meta-Analysis in Environmental Economics.* 1997 ISBN 0-7923-4592-4
13. S. Faucheux, M. O'Connor and J. v.d. Straaten: *Sustainable Development: Concepts, Rationalities and Strategies.* 1998 ISBN 0-7923-4884-2
14. P. Kågeson: *Growth versus the Environment: Is there a Trade-off?.* 1998 ISBN 0-7923-4926-1
15. J.C.J.M. van den Bergh and M.W. Hofkes (eds.): *Theory and Implementation of Economic Models for Sustainable Development.* 1998 ISBN 0-7923-4998-9
16. J.N. Lekakis (ed.): *Freer Trade, Sustainability, and the Primary Production Sector in the Southern EU: Unraveling the Evidence from Greece.* 1998 ISBN 0-7923-5151-7
17. M. Boman, R. Brännlund and B. Kriström (eds.): *Topics in Environmental Economics.* 1999 ISBN 0-7923-5897-x
18. S.M. de Bruyn: *Economic Growth and the Environment.* An Empirical Analysis. 2000 ISBN 0-7923-6153-9
19. C. Kraus: *Import Tariffs as Environmental Policy Instruments.* 2000 ISBN 0-7923-6318-3
20. K. Uno (ed.): *Economy - Energy-Environment Simulation. Beyond The Kyoto Protocol.* 2002 ISBN 1-4020-0450-8

Economy & Environment

21. J. Beghin, D. Roland-Horst and D. Van der Mensbrugghe (eds.): *Trade and the Environment in General Equilibrium: Evidence from Developing Economies*. 2002
ISBN 1-4020-0479-6
22. T. Swanson (ed.): *The Economics of Managing Biotechnologies*. 2002
ISBN 1-4020-0499-0
23. P. Pashardes, T. Swanson and A. Xepapadeas (eds.): *Current Issues in the Economics of Water Resource Management*. Theory, Applications and Policies. 2002
ISBN 1-4020-0542-3

KLUWER ACADEMIC PUBLISHERS – DORDRECHT / BOSTON / LONDON

Printed in the United States
37630LVS00003B/225

9 781402 005428